GCSE Mathematics (9-1)

Edexcel style questions arranged by topic

In the style of:
Pearson Edexcel Level 1/Level 2 GCSE (9 - 1) | **1MA1**

by Peter Bland

© Peter Bland

Contents

Algebra	1
Bearings	23
Bounds	37
Circle Theorems	49
Cumulative Frequency ×3	65
Fractions	90
Frequency	114
Grade 9 1H	140
Grade 9 2H	161
Grade 9 3H	183
Histograms ×3	197
Locus and Constructions	215
Number	234
Probability	256
Quadratics	274
Scatter Graphs	294
Sequences	305
Simultaneous Equations	324
Surds and Indices	335
Transformation of Curves	348
Transformations	359
Trigonometry	377
Vectors	395

Write your name here

Surname

Other names

In the style of:
**Pearson Edexcel
Level 1/Level 2 GCSE (9 - 1)**

Centre Number

Candidate Number

Mathematics
Algebra

Foundation Tier

GCSE style questions arranged by topic

Paper Reference
1MA1/1F

You must have: Ruler graduated in centimetres and millimetres, protractor, pair of compasses, pen, HB pencil, eraser.

Total Marks

Instructions

- Use **black** ink or ball-point pen.
- **Fill in the boxes** at the top of this page with your name, centre number and candidate number.
- Answer **all** questions.
- Answer the questions in the spaces provided
  – there may be more space than you need.
- **Calculators may not be used.**
- Diagrams are **NOT** accurately drawn, unless otherwise indicated.
- You must **show all your working out**.

Information

- The total mark for this paper is
- The marks for **each** question are shown in brackets
  – use this as a guide as to how much time to spend on each question.

Advice

- Read each question carefully before you start to answer it.
- Keep an eye on the time.
- Try to answer every question.
- Check your answers if you have time at the end.

Turn over ▶

© Peter Bland

1 Peter thinks of a number.

   He multiplies the number by 3
   He then adds 2

   His answer is 20
   (a) What number did Peter think of?

   $20 - 2 = 18 \div 3 = 6$

   6
   (2)

   Sophie uses the formula $P = 2a + b$

   to find the perimeter P of this triangle.

   (b) Find the value of P when

   $a = 6$ and $b = 4$

   $P = 2 \times 6 + 4 = 16$

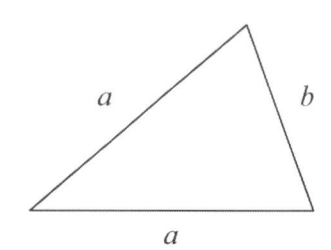

   $P = $ 16
   (2)

   (Total for Question 1 is 4 marks)

2 (a) Work out the value of

   (i) 4^2

   16

   (ii) $\sqrt{64}$

   8

   (iii) 3×2^3

   24
   (3)

   (b) Work out

   (i) $-3 + 5$

   2

   (ii) $-2 - 3$

   .... -5
   (2)

   (Total for Question 2 is 5 marks)

3 The cost of hiring a car can be worked out using this rule.

> Cost = £80 + 50p per mile

Bill hires a car and drives 90 miles.

(a) Work out the cost.

£80 + 50p = 80.50 × 90 = 7245

£ 7245

(2)

The cost of hiring a car and driving m miles is C pounds.

(b) Complete the formula for C in terms of m.

$C =$ ....

(2)

(Total for Question 3 is 4 marks)

4 (a) Complete this table of values for $y = 2x - 1$

x	−1	0	1	2	3	
y		−1		3	5	

(2)

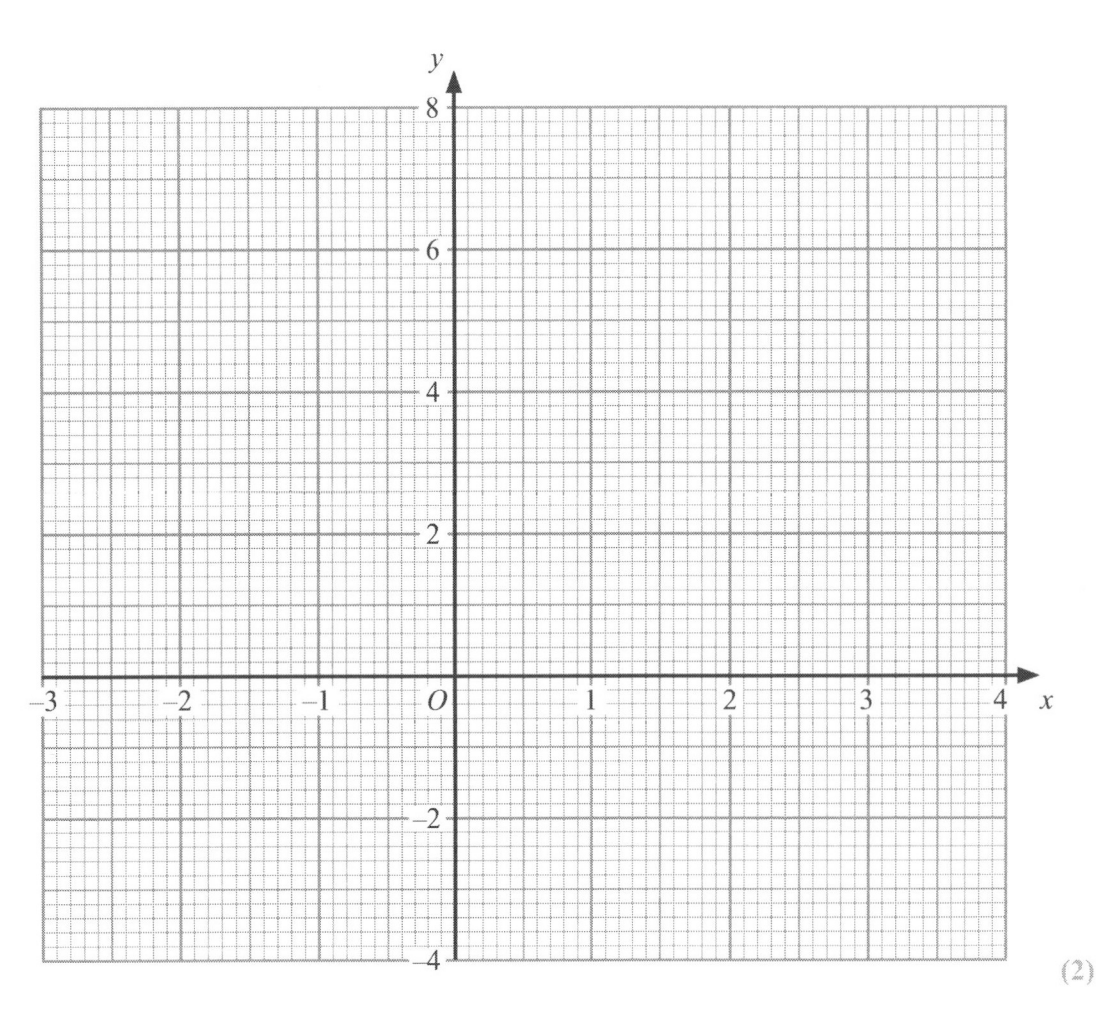

(b) On the grid, draw the graph of $y = 2x - 1$

(2)

(Total for Question 4 is 4 marks)

5 Work out an estimate for the value of $\dfrac{31 \times 4.92}{0.21}$

$\dfrac{30 \times 5}{0} \quad 150$

.... 150 ....

(Total for Question 5 is 2 marks)

6 (a) Expand $y(2y - 3)$

$2y^2 - 3y$

.... $2y^2 - 3y$ ....
(1)

(b) Factorise $x^2 - 4x$

$x(x + 4)$

....
(2)

k is an integer such that $-1 \leqslant k < 3$

(c) List all the possible values of k.

....
(2)

(Total for Question 6 is 5 marks)

7 (a) Factorise $x^2 - 5x$

.... (2)

(b) Expand $3(5x - 2)$

$15x - 6$

$15x - 6$ .... (1)

(Total for Question 7 is 3 marks)

8 A hotel has 64 guests.
40 of the guests are male.

(a) Work out 40 out of 64 as a percentage.

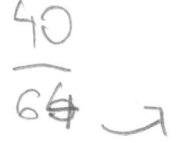

$\dfrac{40}{64}$

.... % (2)

40% of the 40 male guests wear glasses.

(b) Write the number of male guests who wear glasses as a fraction of the 64 guests. Give your answer in its simplest form.

64

×4 (10% = 6.4) ×4
    40% = 25.6

25.6 ....
(2)

(Total for Question 8 is 4 marks)

9 (a) Simplify $8x - 4x$

.... $4x$ ....
(1)

(b) Simplify $y \times y \times y$

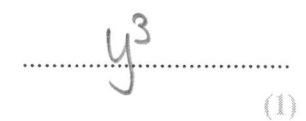

.... y^3 ....
(1)

(c) Simplify $5y + 4x - 2x + 5x$

$5y + 7x$

.... $5y + 7x$ ....
(2)

(Total for Question 9 is 4 marks)

10 The two-way table gives some information about how 100 children travelled to school one day.

	Walk	Car	Bike	Total
Boy	15	25	14	54
Girl	22	8	16	46
Total	37	33	30	100

(a) Complete the two-way table.

(3)

One of the children is picked at random.

(b) Write down the probability that this child walked to school that day.

$\frac{37}{100}$

(1)

One of the girls is picked at random.

(c) Work out the probability that this girl did **not** walk to school that day.

$\frac{24}{100}$

(2)

(Total for Question 10 is 6 marks)

11 Apples cost *a* pence each

Bananas cost *b* pence each.

Write down an expression for the total cost, in pence, of 2 apples and 4 bananas.

.... pence

(Total for Question 11 is 2 marks)

8

12

$4x + 1$

x x

$2x + 12$

Diagram **NOT** accurately drawn

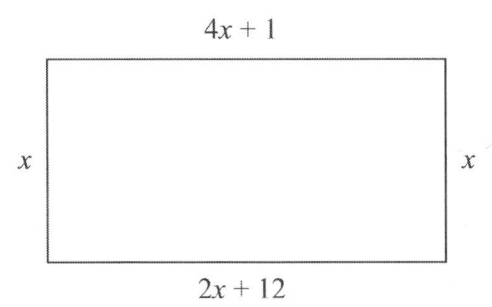

The diagram shows a rectangle.

All the measurements are in centimetres.

(a) Explain why $\quad 4x + 1 = 2x + 12$

....

(1)

(b) Solve $\quad 4x + 1 = 2x + 12$

$4x - 2x = 12 - 1$

$2x = 11$

$x = \frac{11}{2}$

$x = \frac{11}{2}$

(2)

(c) Use your answer to part (b) to work out the perimeter of the rectangle.

.... cm

(1)

(Total for Question 12 is 5 marks)

13 (a) Simplify $5 + 2 - 4cd$

....
(1)

(b) Simplify $\quad 4c + 3d - 2c + 2d$

....
(2)

(c) Simplify $\quad x \times x \times x$

....
(1)

(d) Simplify $\quad 3q \times 2r$

....
(1)

(e) Factorise $\quad 5x + 10$

....
(1)

(Total for question 13 is 6 marks)

14 Expand and simplify $(x + 7)(x + 3)$

....

(Total for Question 14 is 2 marks)

15 Solve $4x + 5 = x + 26$

$x = $ ....

(Total for Question 15 is 2 marks)

16 (a) Tara buys p packets of plain crisps and c packets of cheese crisps.

Write down an expression for the total number of packets of crisps Tara buys.

....

(1)

(b) Solve $3y - 5 = 9$

$y = $ ....

(2)

(Total for Question 16 is 3 marks)

17 Here is a parallelogram.

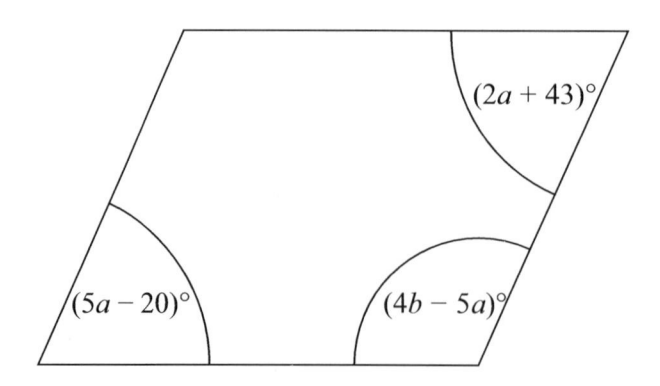

Work out the value of a and the value of b.

Angles shown: $(2a+43)°$, $(5a-20)°$, $(4b-5a)°$

$a = $ ....

$b = $ ....

(Total for Question 17 is 5 marks)

18 (a) Factorise $3f + 9$

....
(1)

(b) Factorise $x^2 - 2x - 15$

....
(2)

(Total for Question 18 is 3 marks)

19 $q = \dfrac{p}{r} + s$

Make p the subject of this formula.

....

(Total for Question 19 is 2 marks)

20 $f = 5x + 2y$

$x = 3$ and $y = -2$

Find the value of f.

....

(Total for Question 20 is 2 marks)

21 Here is a rectangle made of card.

2x
y

The measurements in the diagram are in centimetres.

Sophie fits four of these rectangles together to make a frame.

The perimeter of the inside of the frame is P cm.

(a) Show that $P = 8x - 4y$

(2)

Georgina says,

"When x and y are whole numbers, P is always a multiple of 4."

Is Georgina correct?

You must give a reason for your answer.

....

....

(2)

(Total for Question 21 is 4 marks)

22 You should use a calculator for this question.

The value of a motorhome £V is given by

$$V = 20\,000 \times 0.9^t$$

where t is the age of the motorhome in complete years.

(a) Write down the value of V when $t = 0$.

(a) £ ....
(1)

(b) What is the value of V when $t = 3$?

(b) £ ....
(2)

(c) After how many complete years will the motorhome's value drop below £10 000?

(c) ....
(2)

(Total for Question 22 is 4 marks)

23 Six equations are shown below, each labelled with a letter.

A	B	C
$y = -6x$	$x = \dfrac{1}{6}y$	$y = \dfrac{-3}{x}$

D	E	F
$x = \dfrac{6}{y}$	$y = 6x$	$y = \dfrac{2}{x} + 2$

Choose the correct letters to make each statement true.

(a) Equation B and equation .... are equivalent. (1)

(b) Equation .... and equation .... each show x is inversely proportional to y. (2)

(Total for Question 23 is 3 marks)

---

24 Joe went for a bike ride one evening.

He travelled x kilometres in 5 hours.

Show that his average speed can be written as $\dfrac{x}{18}$ m/s. (4)

(Total for Question 24 is 4 marks)

25 (a) Simplify.

$$x \times x \times x \times x \times x$$

(a) ....
(1)

(b) Solve.

$$3x + 7 = 19$$

(b) $x =$ ....
(2)

(c) Here is a formula.

$$T = 5r + 3u$$

Work out the value of T when $r = 8$ and $u = 9$.

(c) ....
(2)

(Total for Question 25 is 5 marks)

26 (a) Complete this table for $y = 2x - 3$.

x	0	1	2	3	4
y	-3		1		5

(1)

(b) On the grid below, draw the graph of $y = 2x - 3$ for values of x from 0 to 4.

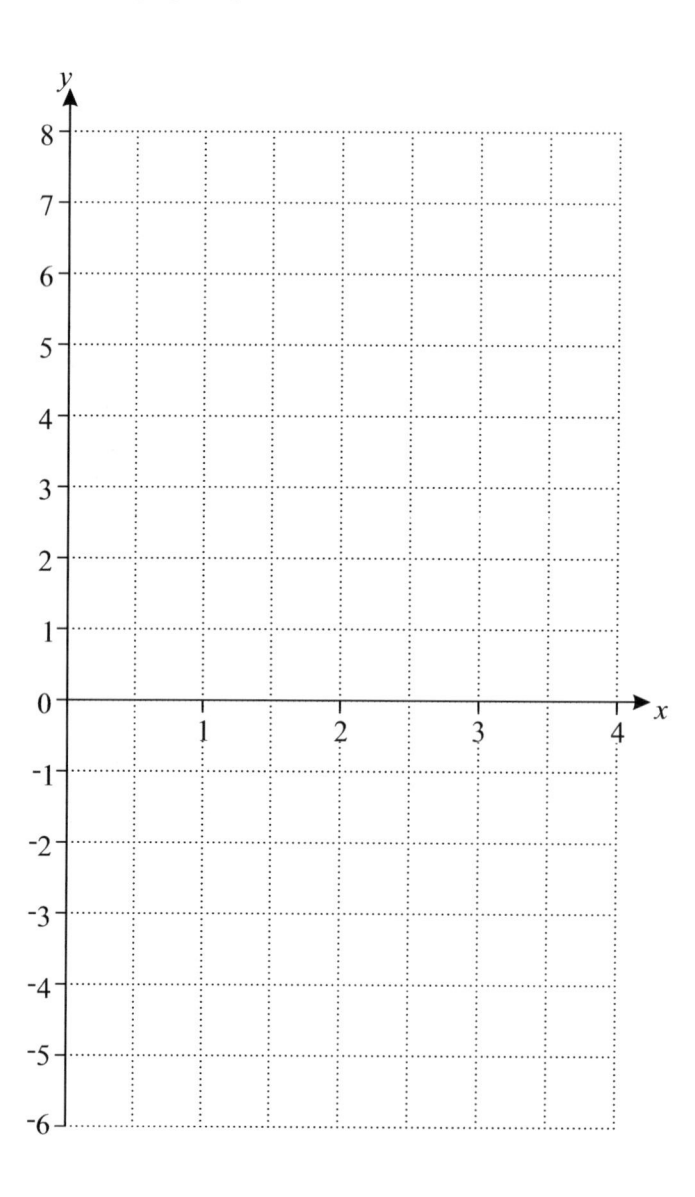

(2)

(c) Line *L* is drawn on the grid below.

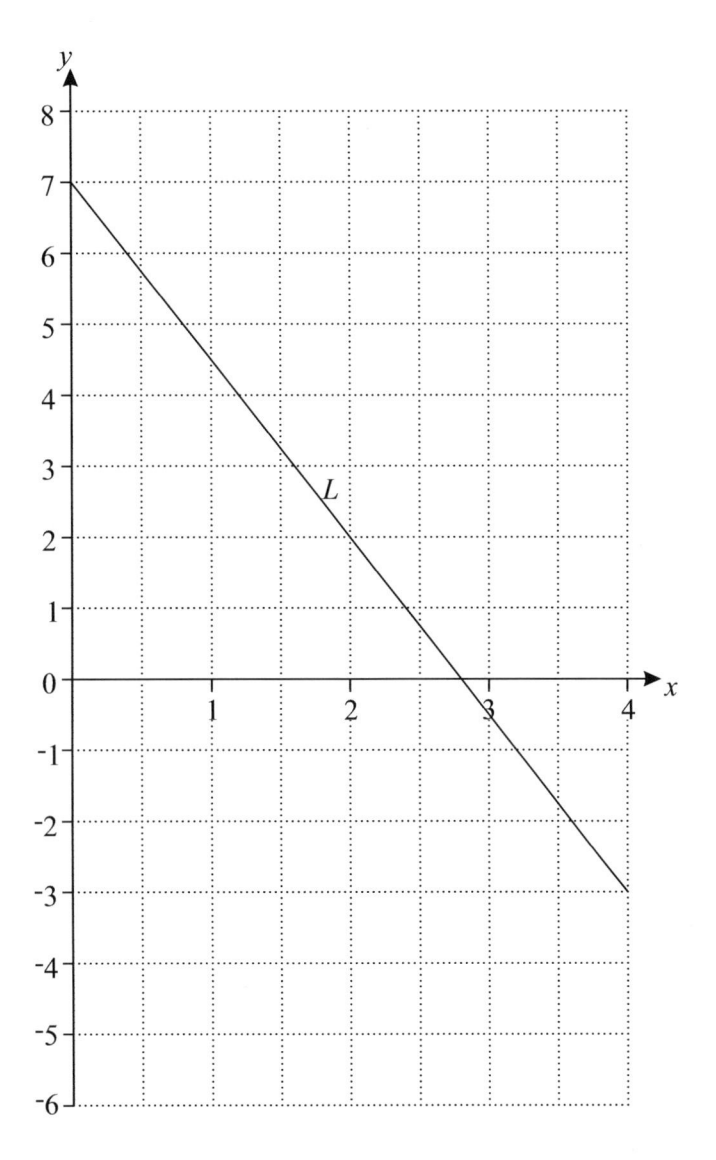

Work out the equation of line *L*.

(c) ....

(3)

(Total for Question 26 is 6 marks)

27

Here is the graph of $y = 5 - x$ for values of x from 0 to 5

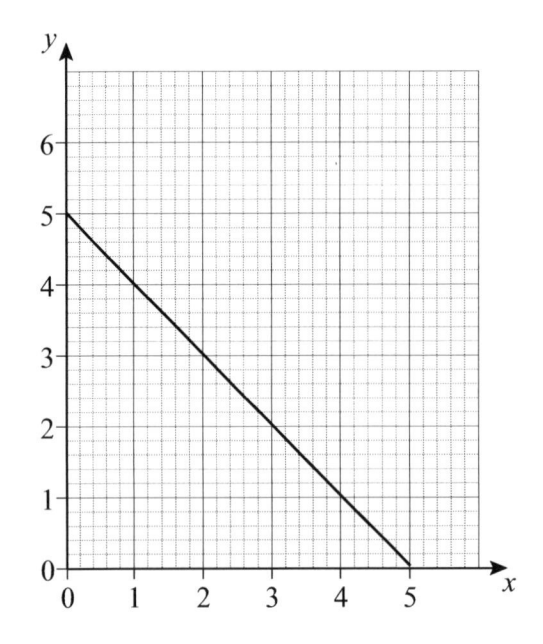

(a) On the same grid, draw the graph of $y = x + 1$ for values of x from 0 to 5

(2)

(b) Use the graphs to solve the simultaneous equations

$y = 5 - x$ and $y = x + 1$

$x = $ ....

$y = $ ....

(1)

(Total for Question 27 is 3 marks)

28 Here are three expressions.

$$\frac{y}{x} \qquad x - yx - y \qquad xy$$

When $x = 2$ and $y = -6$ which expression has the smallest value?

You **must** show your working..

....

(2)

(Total for Question 28 is 2 marks)

---

29 Simplify $5x - (2x + 6)$

Circle your answer.

$3x + 6$ $\qquad$ $9x$ $\qquad$ $-3x$ $\qquad$ $3x - 6$

(Total for Question 29 is 1 mark)

---

30 Helen is trying to work out the two values of w for which $3w - w^3 = 2$

Her values are 1 and -1. Are her values correct?

You **must** show your working.

(2)

(Total for Question 11 is 2 marks)

31

JKLQ is a square.

QLOP is a rectangle.

LMNO is a square.

They are joined to make an L-shape.

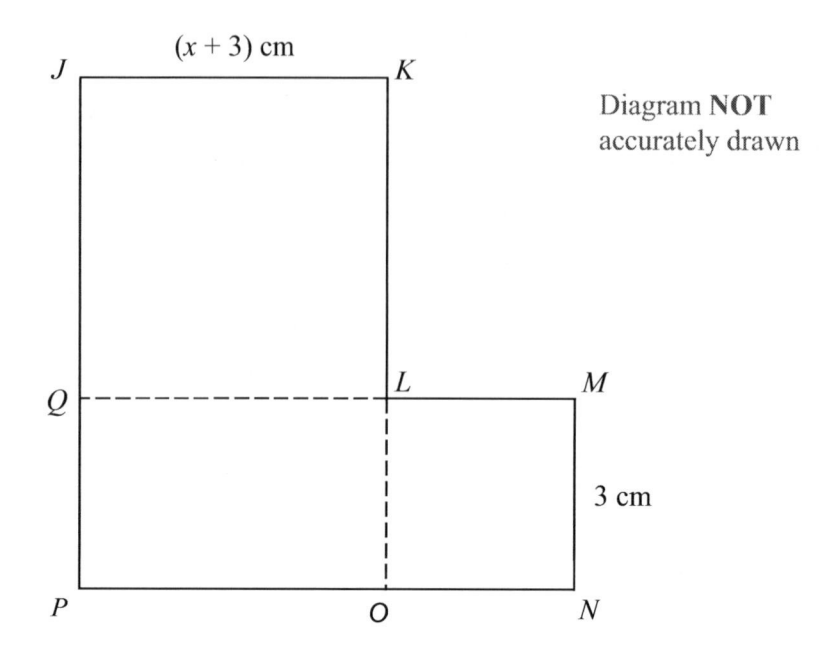

$(x + 3)$ cm

J, K

Diagram **NOT** accurately drawn

Q ----- L, M

3 cm

P, O, N

Show that the total area of the L-shape, in cm², is $x^2 + 9x + 27$

(4)

(Total for Question 31 is 4 marks)

32 Circle the equation with roots 4 and −8

$4x(x - 8) = 0$ $(x - 4)(x + 8) = 0$

$x^2 - 32 = 0$ $(x + 4)(x - 8) = 0$

(Total for Question 32 is 1 mark)

Write your name here

Surname

Other names

In the style of:
Pearson Edexcel
Level 1/Level 2 GCSE (9 - 1)

Centre Number

Candidate Number

Mathematics

Bearings

Foundation Tier

GCSE style questions arranged by topic

Paper Reference
1MA1/2F

You must have: Ruler graduated in centimetres and millimetres, protractor, pair of compasses, pen, HB pencil, eraser, calculator.

Total Marks

Instructions

- Use **black** ink or ball-point pen.
- **Fill in the boxes** at the top of this page with your name, centre number and candidate number.
- Answer **all** questions.
- Answer the questions in the spaces provided
  – *there may be more space than you need.*
- **Calculators may be used.**
- If your calculator does not have a π button, take the value of π to be 3.142 unless the question instructs otherwise.
- Diagrams are **NOT** accurately drawn, unless otherwise indicated.
- You must **show all your working out.**

Information

- The total mark for this paper is
- The marks for **each** question are shown in brackets
  – *use this as a guide as to how much time to spend on each question.*

Advice

- Read each question carefully before you start to answer it.
- Keep an eye on the time.
- Try to answer every question.
- Check your answers if you have time at the end.

Turn over ▶

© Peter Bland

1

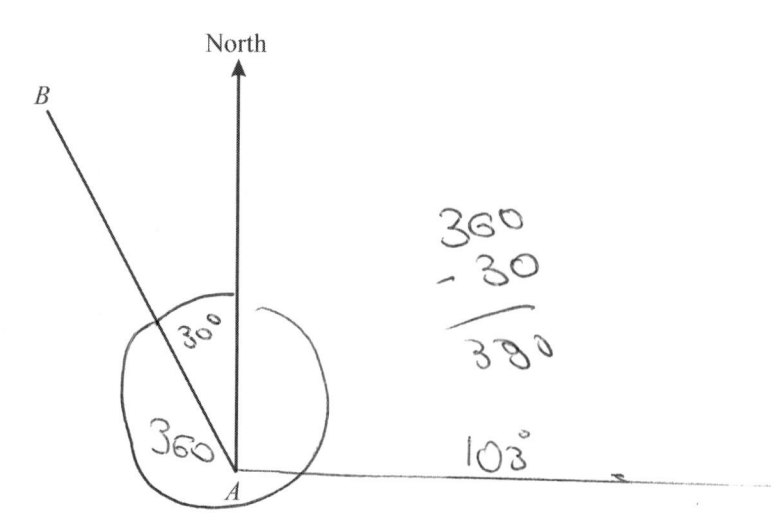

[Diagram: North arrow from A, line to B, with 30° between North and AB, handwritten annotations "30°", "360", and calculation 360 − 30 = 330; also "103°" marked along a line to the right of A]

(a) Measure and write down the bearing of B from A.

    330° .... (1)

(b) On the diagram, draw a line on a bearing of 103° from A.

    (1)

(Total for Question 1 is 2 marks)

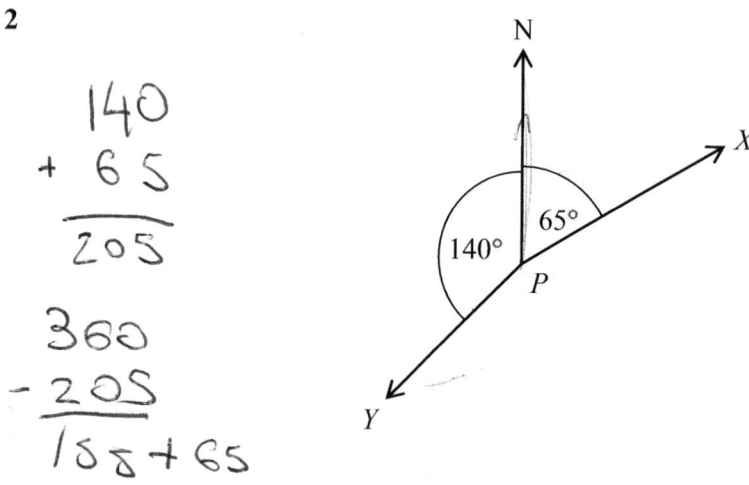

2

[Diagram: point P with North arrow, line to X at 65° from North, angle 140° between the Y-line and North measured round, Y line going to lower left]

Working shown:
140 + 65 = 205
360 − 205 = 155 + 65

(a) Write down the bearing of X from P.

    65° .... (1)

(b) Work out the bearing of Y from P.

    220° ....

3   A ship leaves port X and travels 9 km on a bearing of 120° to point Y.
    The ship then turns and travels 12 km on a bearing of 030° to point Z.
    This journey is shown on the scale drawing below.

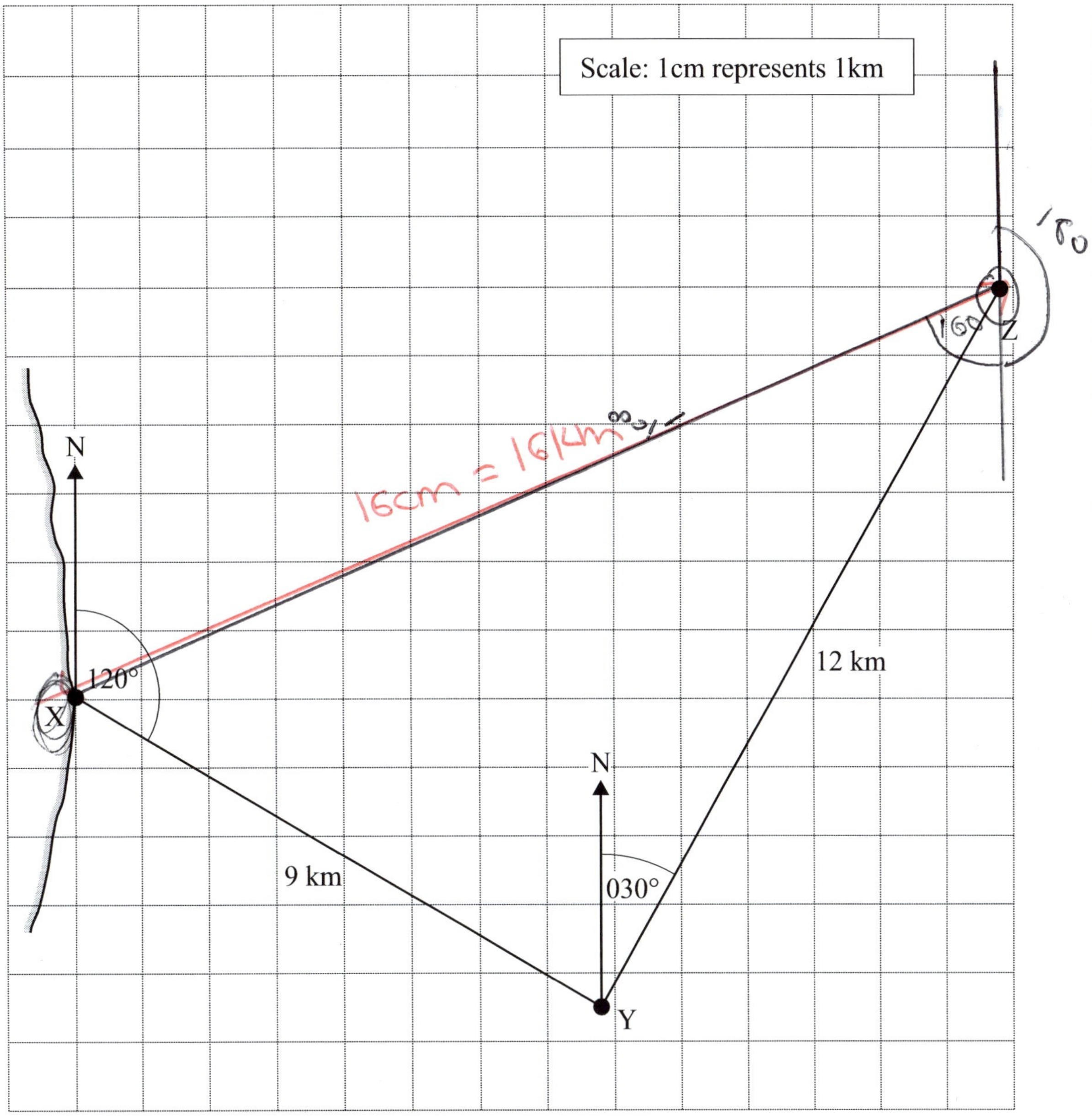

Scale: 1cm represents 1km

[Scale drawing showing ship's journey from X to Y to Z, with handwritten annotations: "16cm = 16km", "180" and "160" near Z, and working "180 + 160"]

The ship then turns and travels directly back from Z to X.

Use a ruler and protractor to work out the distance and bearing of the journey from Z to X

Distance .... 16 km

Bearing .... 340 °

(3)

(Total for Question 3 is 3 marks)

4   An helicopter flies due North from X to Y.
    The distance from X to Y on the river is 24 miles.

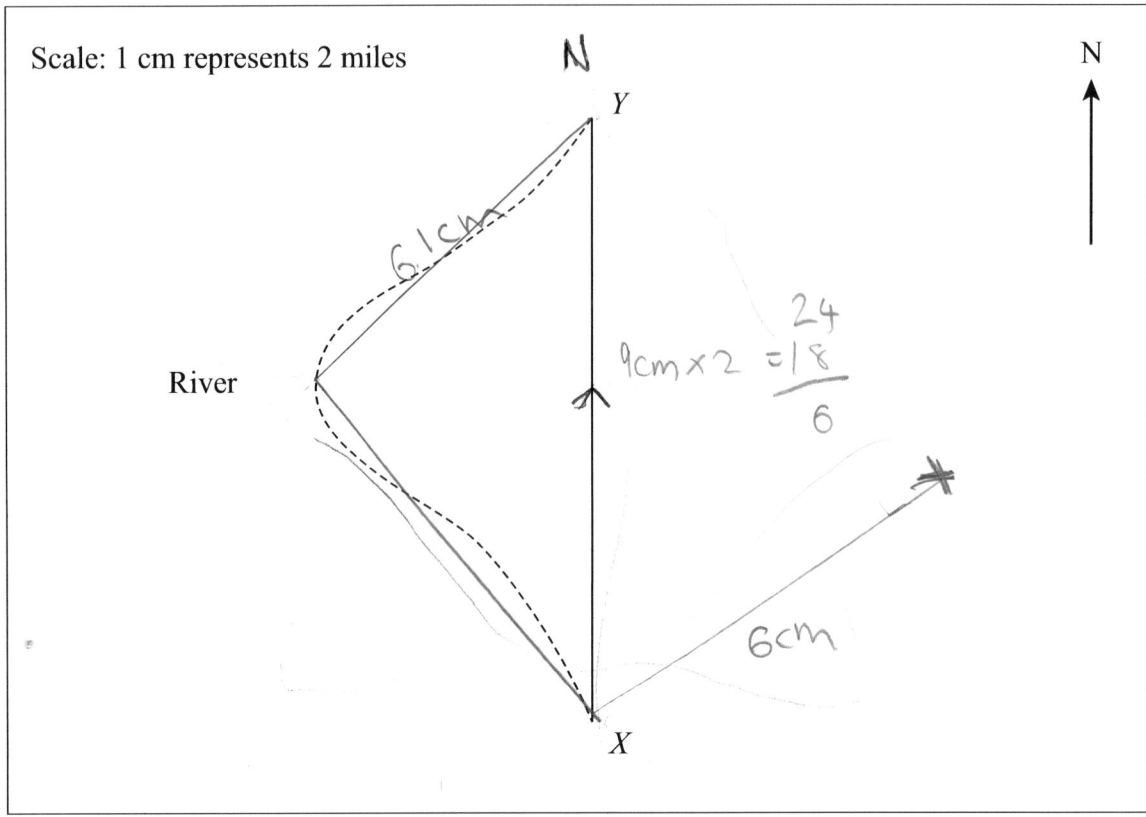

Scale: 1 cm represents 2 miles

River

handwritten annotations on diagram: 6.1cm, 9cm × 2 = 18, 24/6, 6cm

Y
X

(a) How much further is it from X to Y on the river than by helicopter?

.............6...... miles
(3)

(b) Z is 12 miles north-east of X.

(i) Write down the three-figure bearing of Z from X.

.............43........ °
(1)

(ii) Mark with a cross the point Z on the diagram.

(2)

(Total for Question 4 is 6 marks)

5 The diagram shows the positions of two telephone masts, X and Y, on a map.

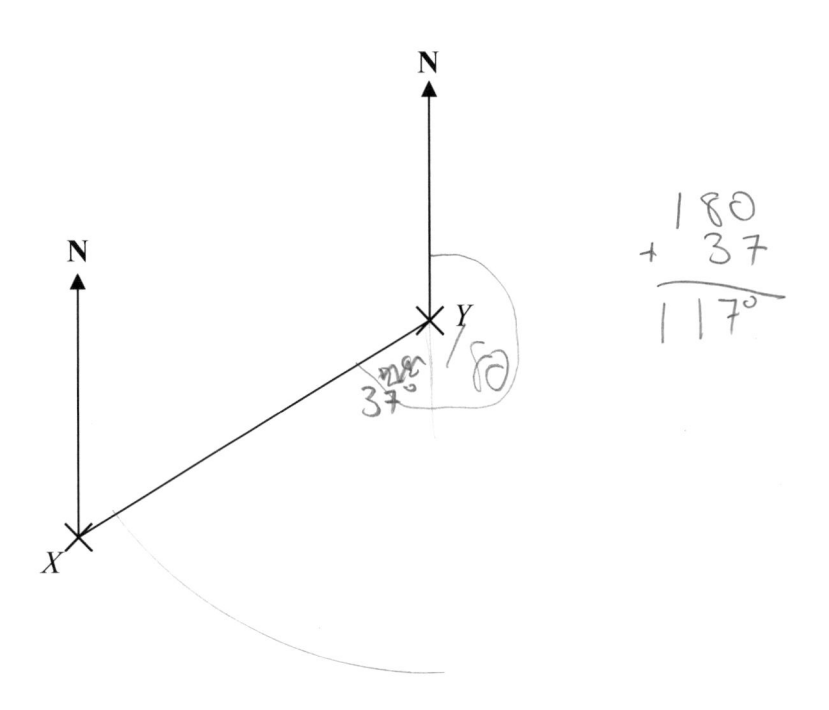

(a) Measure the bearing of Y from X. ....117............°
(1)

Another mast Z is on a bearing of 160 from Y.

On the map, Z is 4 cm from Y.

(b) Mark the position of Z with a cross (X) and label it Z.
(2)

(Total for Question 5 is 3 marks)

6   The diagram shows part of a map.
     It shows the positions of a lighthouse and a boat.

     lighthouse ✗

     ✗ boat

     The scale of the map is 1:10 000

     (a) Work out the real distance between the lighthouse and the boat.
         Give your answer in metres.

     .... m   (2)

     (b) Find the bearing of the lighthouse from the boat.

     ....
                                              (1)

                                    (Total for Question 6 is 3 marks)

7 The diagram shows the position of two ports, *A* and *B*.
A ship sails from port *A* to port *B*.

Scale: 1 cm represents 50 km

(a) Measure the size of the angle marked *x*.

.... °
(1)

(b) Work out the real distance between port *A* and port *B*.
Use the scale 1 cm represents 50 km.

.... km
(2)

Port *C* is 350 km on a bearing of 060° from port *B*.

(c) On the diagram, mark airport *C* with a cross (×).
Label it *C*.

(2)

(Total for Question 7 is 5 marks)

8 Peter keeps bees in two beehives.
They are marked P and Q in the scale drawing below.

Scale: 1 cm represents 50 metres

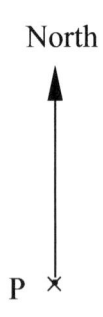

North

P ×

× Q

(a) If Peter walks at about 2 m/s, estimate how long it takes him to walk from beehive P to beehive Q.

(a) ....
(3)

(b) Bees can indicate to other bees where flowers are.

A bee indicates that there are flowers

- on a bearing of 055° from P
- at a distance of 400 m from P.

On the scale drawing, show the point where the flowers are.
Label this point F.

(2)

(c) Peter plants some fruit trees, which are

- the same distance from P and from Q
- 200 m or less from P.

Indicate on the scale drawing where Peter plants the trees.
You must show all your construction lines.

(4)

(Total for Question 8 is 9 marks)

9 The diagram shows the positions of a tower and a tree.

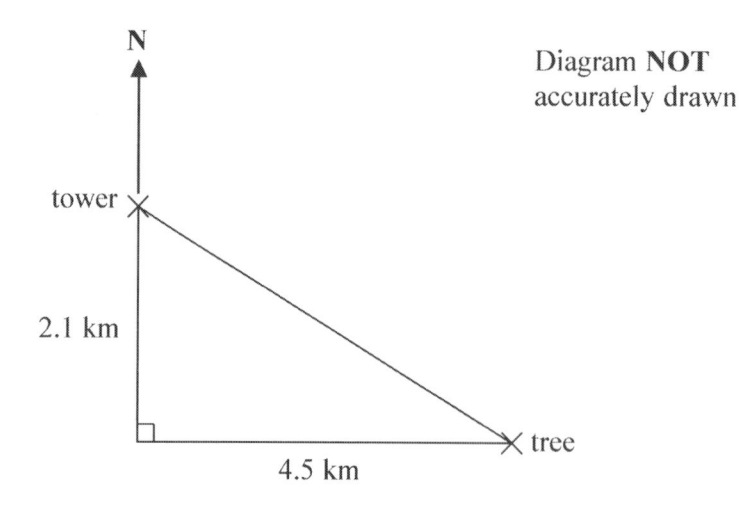

Diagram **NOT** accurately drawn

The tree is 2.1 km South of the tower and 4.5 km East of the tower.

(a) Work out the distance between the tower and the tree.
    Give your answer correct to one decimal place.

(3)

(b) Work out the bearing of the tree from the tower.
    Give your answer correct to the nearest degree.

(4)

(Total for Question 9 is 7 marks)

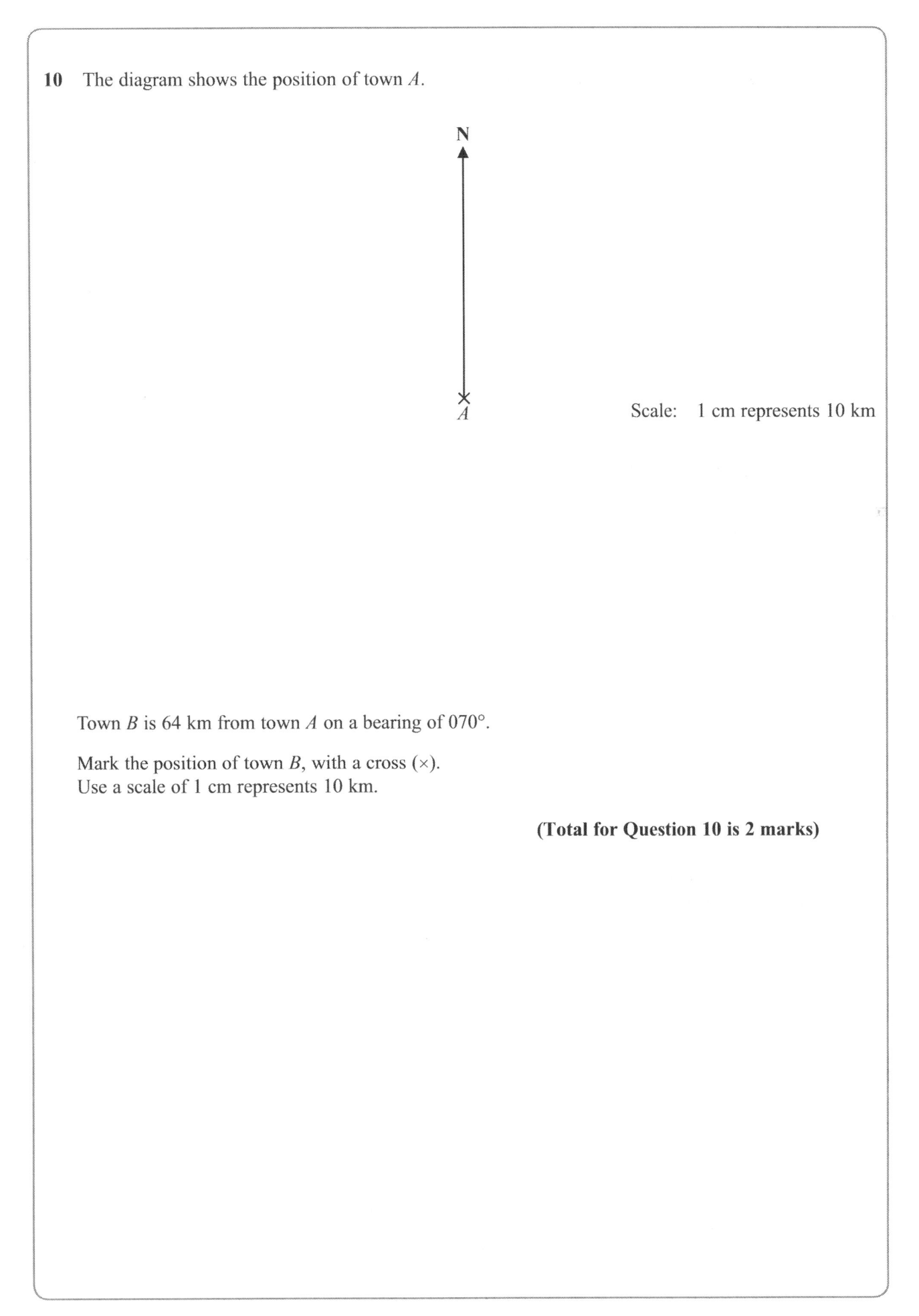

10   The diagram shows the position of town *A*.

Scale: 1 cm represents 10 km

Town *B* is 64 km from town *A* on a bearing of 070°.

Mark the position of town *B*, with a cross (×).
Use a scale of 1 cm represents 10 km.

(Total for Question 10 is 2 marks)

11 The diagram shows the position of two boats, *B* and *C*.

Boat *T* is on a bearing of 060° from boat *B*.
Boat *T* is on a bearing of 285° from boat *C*.

In the space above, draw an accurate diagram to show the position of boat *T*.

Mark the position of boat *T* with a cross (×).
Label it *T*.

(Total for Question 11 is 3 marks)

12 The diagram shows an island with North lines drawn at points A and B.

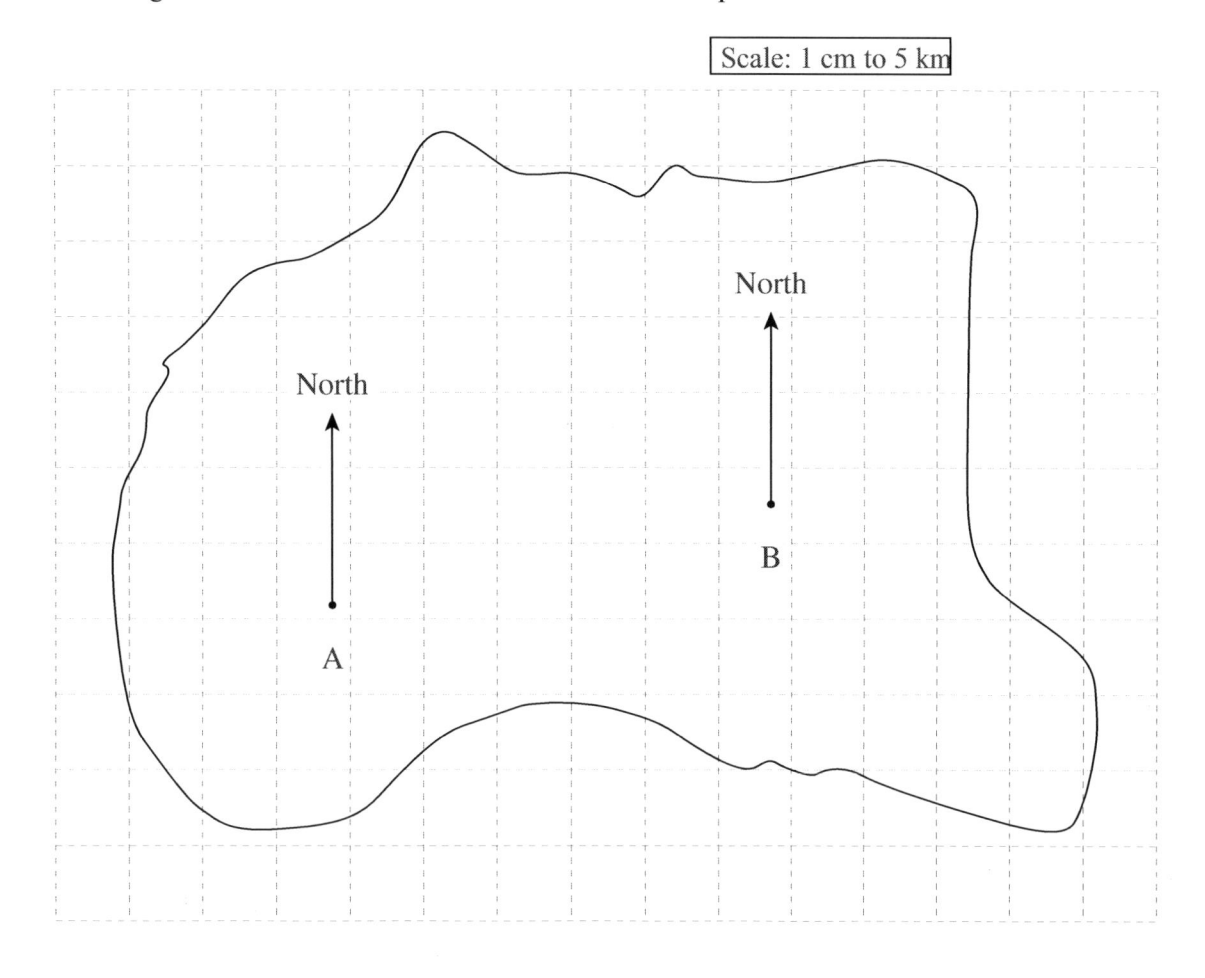

Scale: 1 cm to 5 km

(a) Treasure is buried on a bearing of 037° from A and 290° from B.
Mark, with a ×, the position of the treasure.

(3)

(b) Find the real distance between the points A and B.

.... km

(3)

(Total for Question 12 is 6 marks)

13   The map of an island is shown.

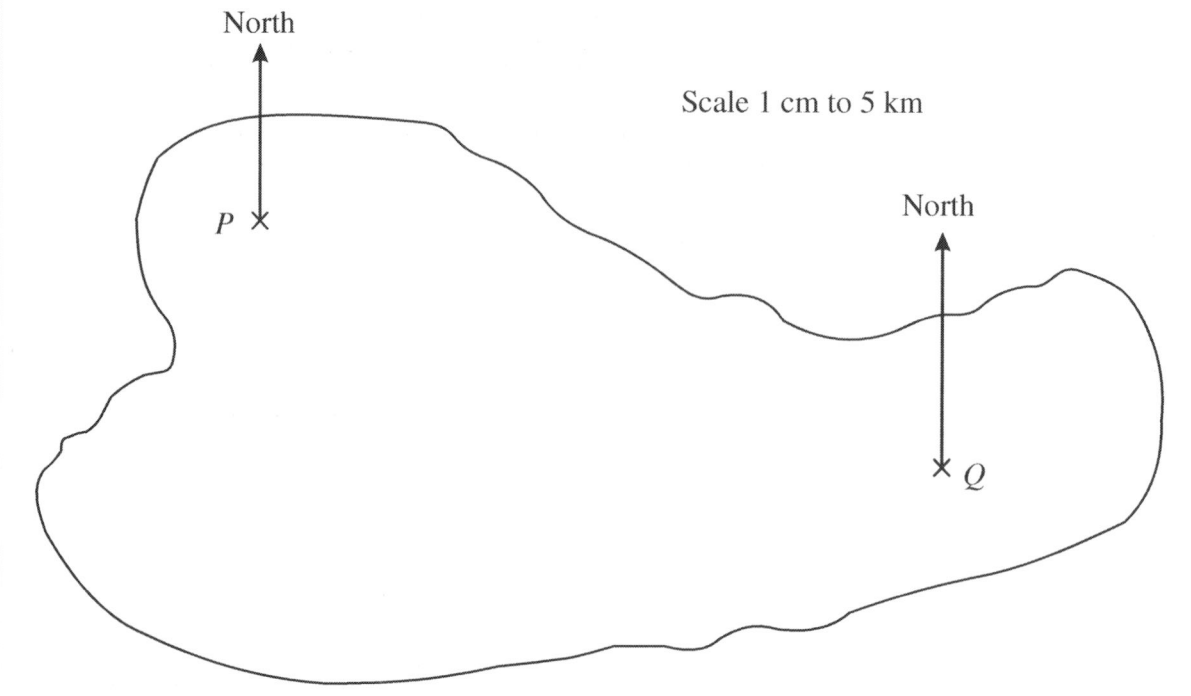

Scale 1 cm to 5 km

P and *Q* are the positions of two houses on the island.

(a)   What is the bearing of *P* from *Q*?

....°

(1)

(b)   Calculate the actual distance from *P* to *Q* in kilometres.

....km

(2)

(c)   A house is 20 km from *P* on a bearing of 130°.
      Mark the position of the house on the diagram with an **X**.

(2)

(Total for Question 13 is 5 marks)

Write your name here

Surname

Other names

In the style of:
Pearson Edexcel
Level 1/Level 2 GCSE (9 - 1)

Centre Number

Candidate Number

Mathematics
Bounds

Higher Tier

GCSE style questions arranged by topic

Paper Reference
1MA1/2H

You must have: Ruler graduated in centimetres and millimetres, protractor, pair of compasses, pen, HB pencil, eraser, calculator.

Total Marks

Instructions

- Use **black** ink or ball-point pen.
- **Fill in the boxes** at the top of this page with your name, centre number and candidate number.
- Answer **all** questions.
- Answer the questions in the spaces provided
  – *there may be more space than you need.*
- **Calculators may be used.**
- If your calculator does not have a π button, take the value of π to be 3.142 unless the question instructs otherwise.
- Diagrams are **NOT** accurately drawn, unless otherwise indicated.
- You must **show all your working out**.

Information

- The total mark for this paper is
- The marks for **each** question are shown in brackets
  – *use this as a guide as to how much time to spend on each question.*

Advice

- Read each question carefully before you start to answer it.
- Keep an eye on the time.
- Try to answer every question.
- Check your answers if you have time at the end.

Turn over ▶

© Peter Bland

1   $w = \sqrt{\dfrac{x}{y}}$

$x = 5.43$ correct to 2 decimal places.

$y = 4.514$ correct to 3 decimal places.

By considering bounds, work out the value of w to a suitable degree of accuracy.

You must show all your working and give a reason for your final answer.

$w =$ ....

(Total for Question 1 is 5 marks)

2 An arrow is shot vertically upwards at a speed of V metres per second.

The height, H metres, to which it rises is given by

$$H = \frac{V^2}{2g}$$

where g m/s² is the acceleration due to gravity.

$V = 24.4$ correct to 3 significant figures.

$g = 9.8$ correct to 2 significant figures.

(i) Write down the upper bound of g.

....
(1)

(ii) Calculate the lower bound of H.
Give your answer correct to 3 significant figures.

....
(2)

(Total for Question 2 is 3 marks)

3 A building plot is in the shape of a rectangle.
The width of the field is 26 metres, measured to the nearest metre.

(a) Work out the upper bound of the width of the field.

.... metres
(1)

The length of the field is 135 metres, measured to the nearest 5 metres.

(b) Work out the upper bound for the perimeter of the field.

.... metres
(3)

(Total for Question 3 is 4 marks)

4 Sophie drove for 238 miles, correct to the nearest mile.
   She used 26.3 litres of petrol, to the nearest tenth of a litre.

$$\text{Petrol consumption} = \frac{\text{Number of miles travelled}}{\text{Number of litres of petrol used}}$$

Work out the upper bound for the petrol consumption for Sophie's journey. Give your answer correct to 2 decimal places.

.... miles per litre

(Total for Question 4 is 3 marks)

5 (a) A solid cube has sides of length 5 cm.

Diagram **NOT** accurately drawn

Work out the total surface area of the cube. State the units of your answer.

....

(4)

(b) Change 125 cm³ into mm³.

.... mm³

(2)

The weight of the cube is 77 grams, correct to the nearest gram.

(c) (i) What is the minimum the weight could be?

.... grams

(ii) What is the maximum the weight could be?

.... grams

(2)

(Total for Question 5 is 8 marks)

6 The length of a line is 53 centimetres, correct to the nearest centimetre.

(a) Write down the **least** possible length of the line.

.... centimetres
(1)

(b) Write down the **greatest** possible length of the line.

.... centimetres
(1)

(Total for Question 6 is 2 marks)

7 The voltage V of an electronic circuit is given by the formula

$$V = IR$$

where I is the current in amps
and R is the resistance in ohms.

Given that $\quad V = 208 \quad$ correct to 3 significant figures,
$\quad\quad\quad\quad\quad\quad R = 12.8 \quad$ correct to 3 significant figures,

calculate the lower bound of I.

....

(Total for Question 7 is 3 marks)

8 The average fuel consumption (*c*) of Tara's car, in kilometres per litre, is given by the formula

$$c = \frac{d}{f}$$

where *d* is the distance travelled, in kilometres, and *f* is the fuel used, in litres.

$d = 153$ correct to 3 significant figures.
$f = 43.3$ correct to 3 significant figures.

By considering bounds, work out the value of *c* to a suitable degree of accuracy.
You must show **all** of your working **and** give a reason for your final answer.

c = ...

(Total for Question 8 is 5 marks)

9 A box is a cuboid with dimensions 27 cm by 15 cm by 20 cm

These dimensions are to the nearest **centimetre**.

DVD cases are cuboids with dimensions 1.5 cm by 14.3 cm by 19.2 cm

These dimensions are to the nearest **millimetre**.

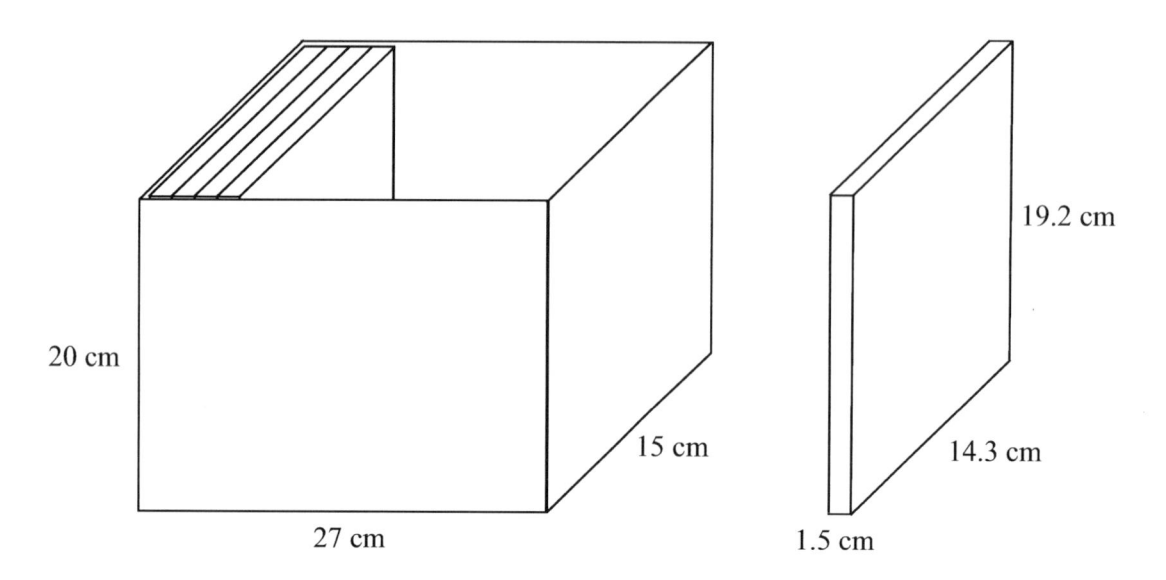

Show that 17 DVD cases, stacked as shown, will definitely fit in the box.

(Total for Question 9 is 4 marks)

10     $m = \dfrac{\sqrt{s}}{t}$ $s = 3.47$ correct to 3 significant figures

                       $t = 8.132$ correct to 4 significant figures

By considering bounds, work out the value of m to a suitable degree of accuracy.
Give a reason for your answer.

(Total for Question 10 is 5 marks)

11 (a) The attendance at a football match was 67 500, correct to the nearest hundred.

(i) What was the **highest** possible attendance?

(a)(i) ....
(1)

(ii) What was the **lowest** possible attendance?

(ii) ....
(1)

(b) A distance, *d*, was given as 6.73 m, **truncated** to 2 decimal places.

Complete the error interval for the distance, *d*.

(2)

.... ≤ *d* < ....

(Total for Question 11 is 4 marks)

Write your name here

Surname

Other names

In the style of:
Pearson Edexcel
Level 1/Level 2 GCSE (9 - 1)

Centre Number

Candidate Number

Mathematics

Circle Theorems

Higher Tier

GCSE style questions arranged by topic

Paper Reference
1MA1/2H

You must have: Ruler graduated in centimetres and millimetres, protractor, pair of compasses, pen, HB pencil, eraser, calculator.

Total Marks

Instructions

- Use **black** ink or ball-point pen.
- **Fill in the boxes** at the top of this page with your name, centre number and candidate number.
- Answer **all** questions.
- Answer the questions in the spaces provided
  – *there may be more space than you need.*
- **Calculators may be used.**
- If your calculator does not have a π button, take the value of π to be 3.142 unless the question instructs otherwise.
- Diagrams are **NOT** accurately drawn, unless otherwise indicated.
- You must **show all your working out**.

Information

- The total mark for this paper is
- The marks for **each** question are shown in brackets
  – *use this as a guide as to how much time to spend on each question.*

Advice

- Read each question carefully before you start to answer it.
- Keep an eye on the time.
- Try to answer every question.
- Check your answers if you have time at the end.

Turn over ▶

© Peter Bland

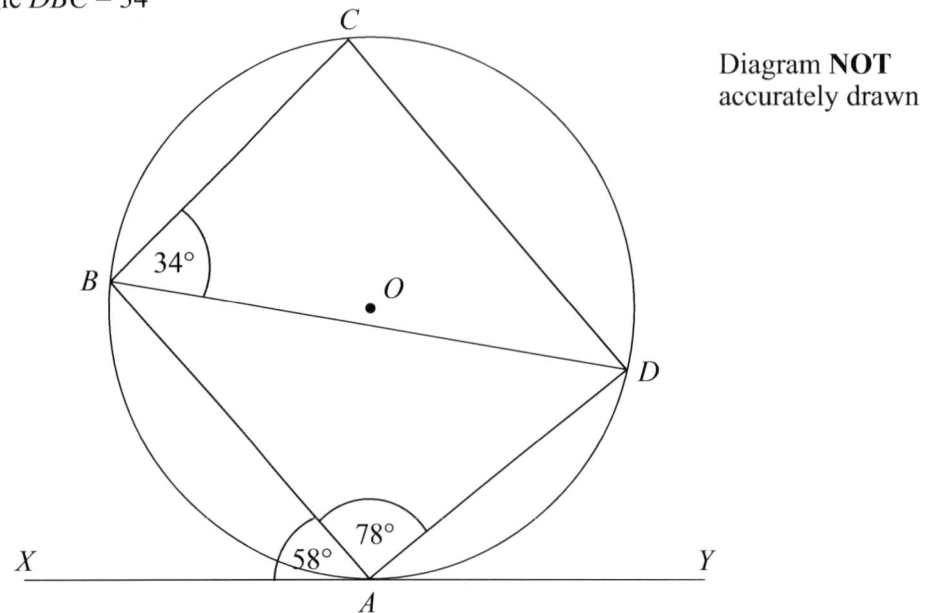

1. *ABCD* is a cyclic quadrilateral within a circle centre *O*.
   XY is the tangent to the circle at *A*.
   Angle *XAB* = 58°
   Angle *BAD* = 78°
   Angle *DBC* = 34°

   Diagram **NOT** accurately drawn

   Prove that *AB* is parallel to *CD*.

   (Total for Question 1 is 5 marks)

2 (a) Here is a circle with centre O.

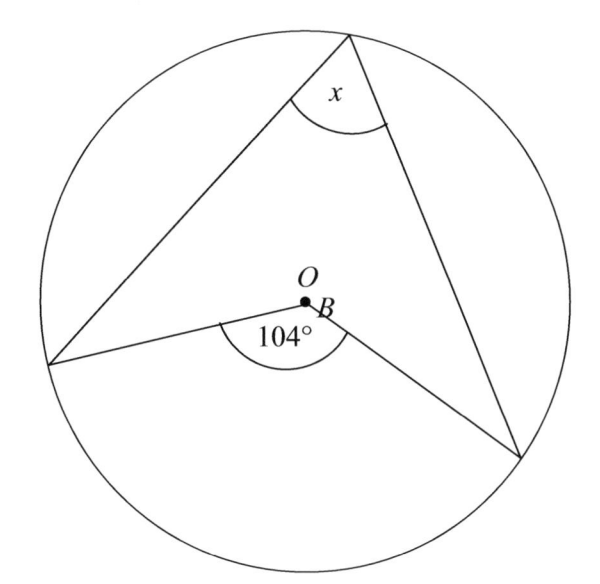

Diagram **NOT** accurately drawn

Write down the value of x.

.... degrees
(1)

(b) Here is a different circle.

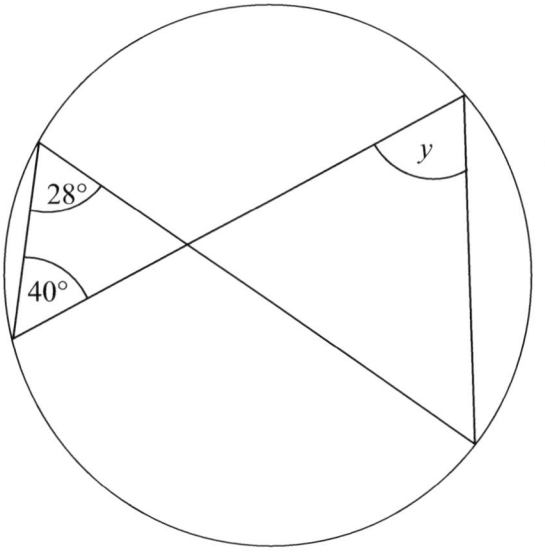

Diagram **NOT** accurately drawn

Write down the value of y.

.... degrees
(1)

(Total for Question 2 is 2 marks)

3

U, *V* and *W* are points on the circumference of a circle, centre *O*. *UW* is a diameter of the circle.

(a) (i) Write down the size of angle *UVW*.

.... °

(ii) Give a reason for your answer.

....

....

(2)

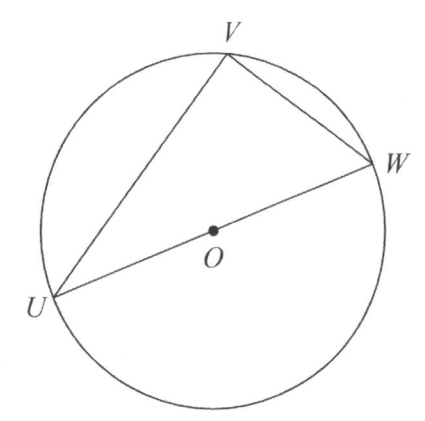

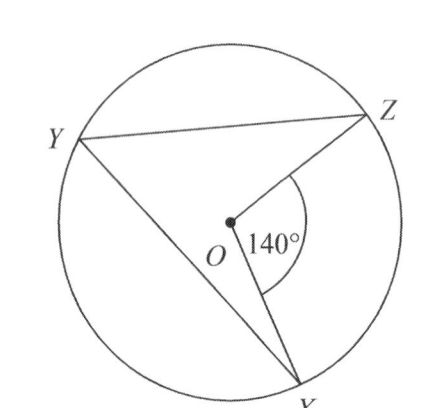

Diagram **NOT** accurately drawn

X, *Y* and *Z* are points on the circumference of a circle, centre *O*.
Angle *XOZ* = 140°.

(b) (i) Work out the size of angle *XYZ*.

.... °

(ii) Give a reason for your answer.

....

....

(2)

(Total for Question 3 is 4 marks)

4.

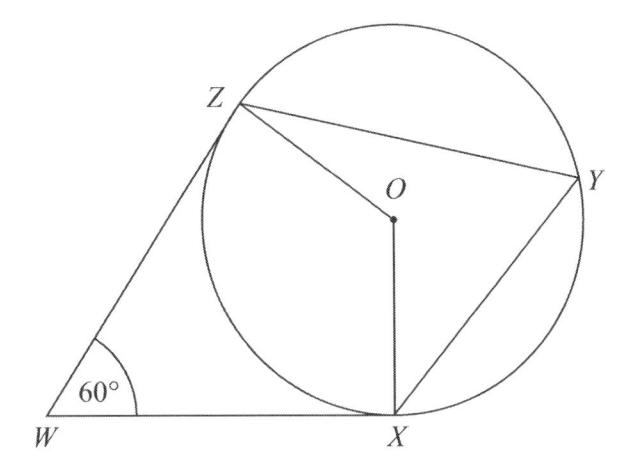

Diagram **NOT** accurately drawn

X, Y and Z are points on the circumference of a circle, centre O.
WX and WZ are tangents to the circle.

Angle $ZWX = 60°$

Work out the size of angle XYZ.
Give a reason for each stage in your working.

(Total for Question 4 is 4 marks)

5

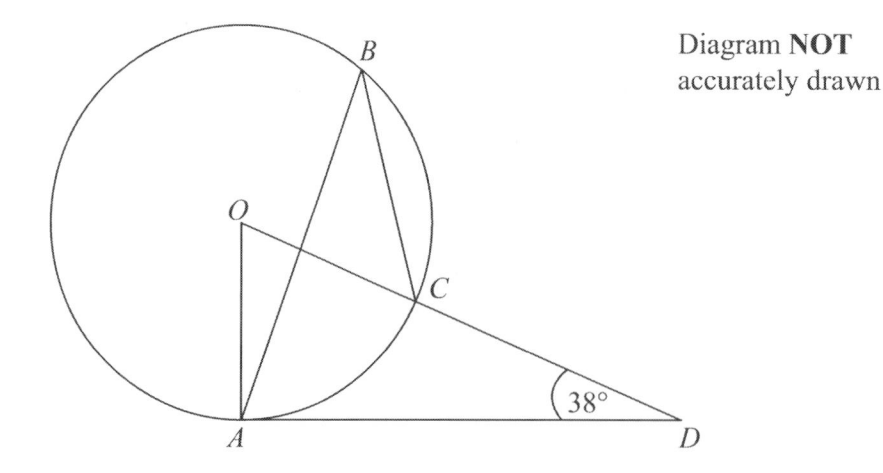

Diagram **NOT** accurately drawn

The diagram shows a circle centre O.
A, B and C are points on the circumference.

DCO is a straight line.
DA is a tangent to the circle.

Angle $ADO = 38°$

(a) Work out the size of angle AOD.

.... °

(2)

(b) (i) Work out the size of angle ABC.

.... °

(ii) Give a reason for your answer.

....

(3)

(Total for Question 5 is 5 marks)

6

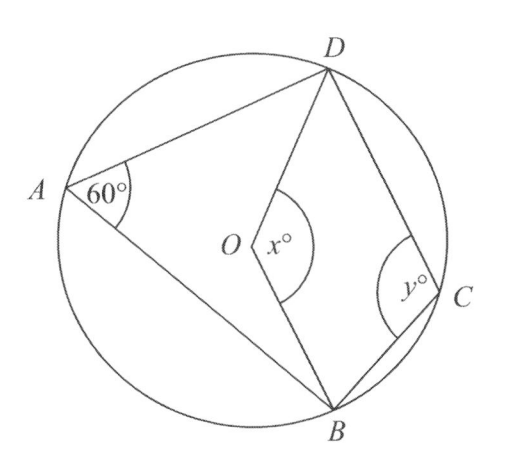

Diagram **NOT** accurately drawn

In the diagram, A, B, C and D are points on the circumference of a circle, centre O.
Angle BAD = 60°.
Angle BOD = x°.
Angle BCD = y°.

(a) (i) Work out the value of x.

x = ...

(ii) Give a reason for your answer.

....

....

(2)

(b) (i) Work out the value of y.

y = ...

(ii) Give a reason for your answer.

....

....

(2)

(Total for Question 6 is 4 marks)

7

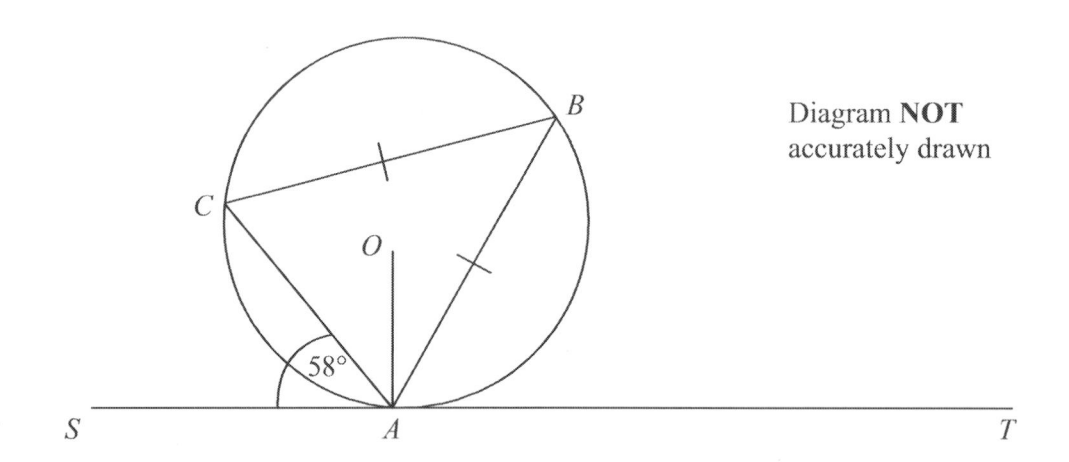

Diagram **NOT** accurately drawn

A, B and C are points on the circumference of a circle, centre O.
The line SAT is the tangent at A to the circle.

$CB = AB$.
Angle $CAS = 58°$.

Calculate the size of angle OAB.
Give a reason for each stage in your working.

.... °

(Total for Question 7 is 5 marks)

8

A, *B*, *C* and *D* are points on the circumference of a circle.
Angle *ABD* = 54°.
Angle *BAC* = 28°.

(i) Find the size of angle *ACD*.

.......................°

(ii) Give a reason for your answer.

....

....

(Total for Question 8 is 2 marks)

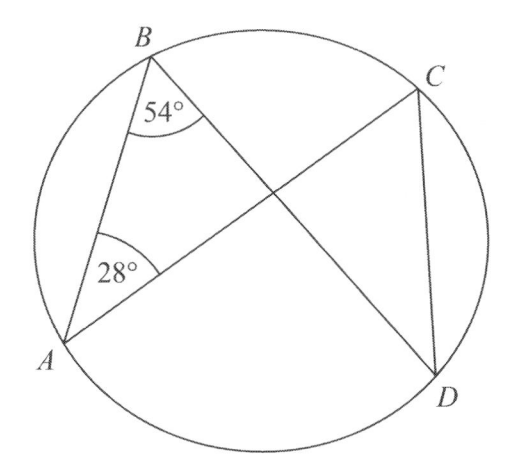

9

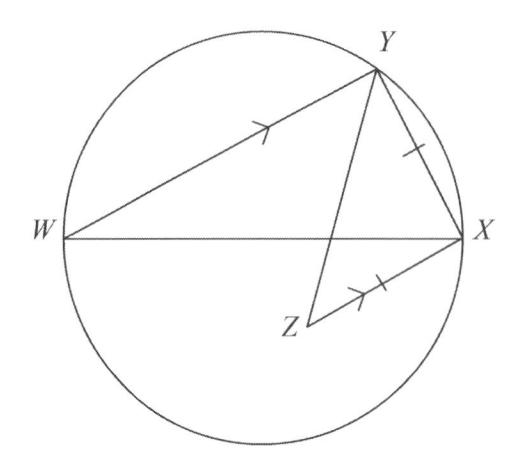

Diagram **NOT** accurately drawn

WX is a diameter of a circle.
Y is a point on the circle.
Z is the point inside the circle such that ZX = XY and XZ is parallel to YW.
Find the size of angle XZY.
You must give reasons for your answer.

....................................°

(Total for Question 9 is 4 marks)

10. *ABCD* is a cyclic quadrilateral.
AE is a tangent at *A*.
CDE is a straight line.
Angle *CAD* = 32°
Angle *ABD* = 40°

Diagram **NOT** accurately drawn

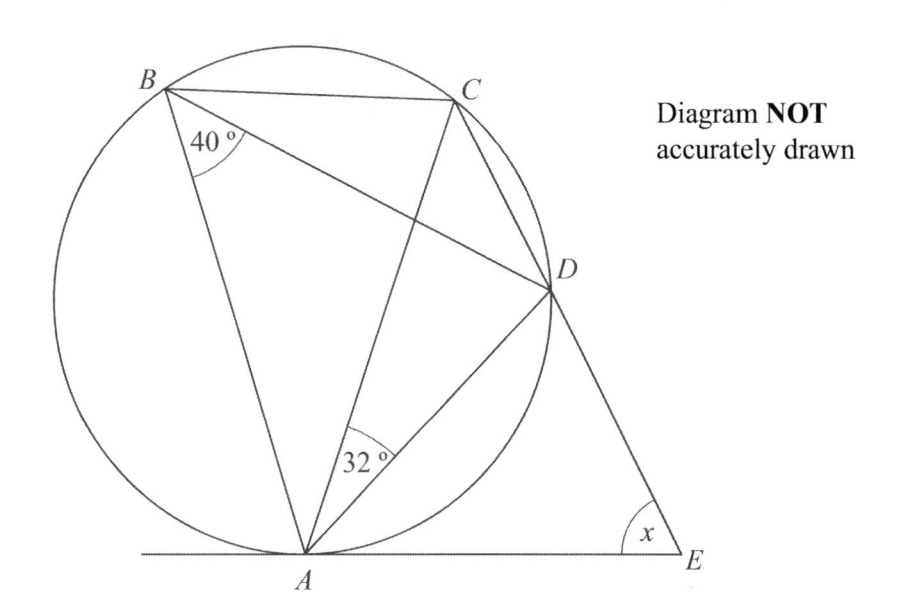

Work out the size of angle *AED*, marked *x*, on the diagram.
You **must** show your working.
Give reasons for any angles you work out.

.... degrees

(Total for Question 10 is 5 marks)

11

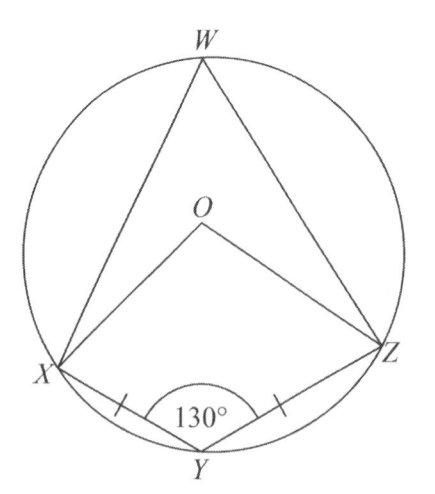

Diagram NOT accurately drawn

W, X, Y and Z are points on a circle, centre O.
XY = YZ.
Angle XYZ = 130°.

(a) Write down the size of angle XWZ.
    Give a reason for your answer.

.................................°

(2)

(b) Work out the size of angle OZY.
    Give reasons for your answer.

.................................°

(4)

(Total for Question 11 is 6 marks)

12 *A, B, C* and *D* are points on the circumference of a circle, centre *O*.
    AC is a diameter of the circle.
    Angle *ABD* = 58°.
    Angle *CDB* = 22°.

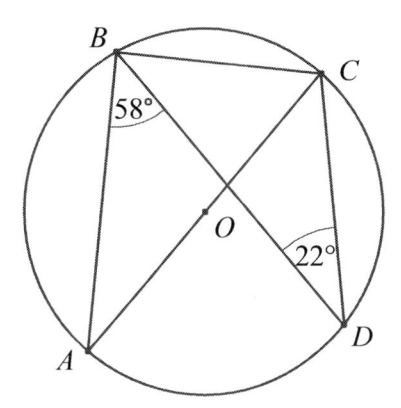

    Diagram **NOT** accurately drawn

    Work out the sizes of angle *ACD* and *ACB*, giving reasons for your answers.

    (a) Angle *ACD* = ...°

    (2)

    (b) Angle *ACB* = ...°

    (3)

    (Total for Question 12 is 5 marks)

Points *P, Q, R* and *S* lie on the circumference of the circle.
UST is a tangent to the circle.
Angle *RPS* = 44° and angle *PSO* = 32°.

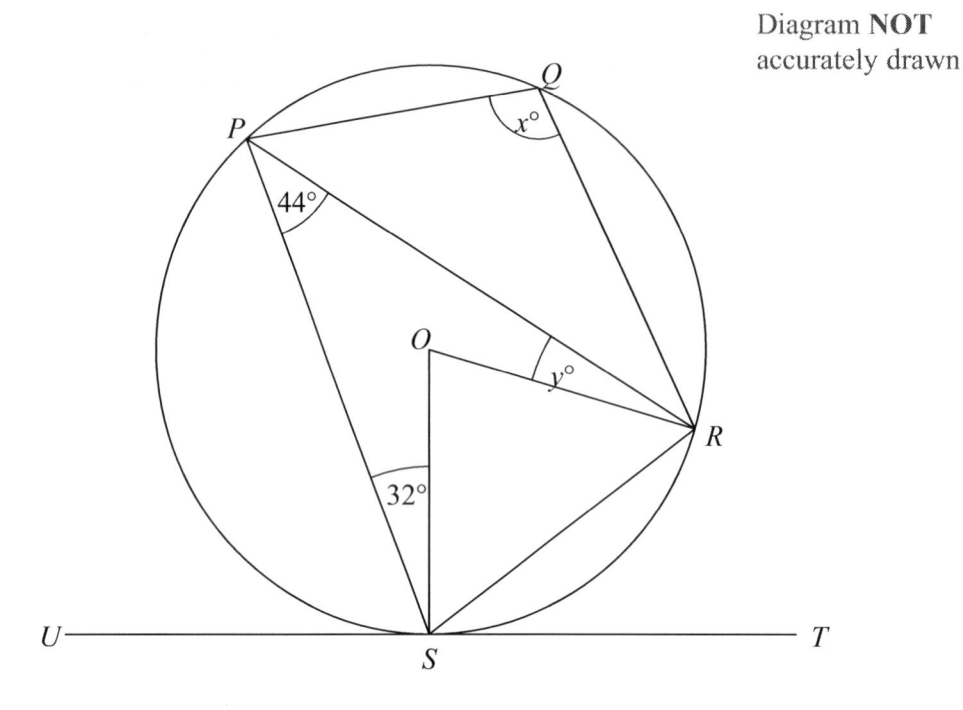

Diagram **NOT** accurately drawn

(a) Work out the value of *x*.

x = ...
(4)

(b) Work out the value of *y*.

y = ...

(Total for Question 13 is 7 marks)
(3)

14   *A*, *B* and *C* are points on a circle.

- *BC* bisects angle *ABQ*.
- *PBQ* is a tangent to the circle.

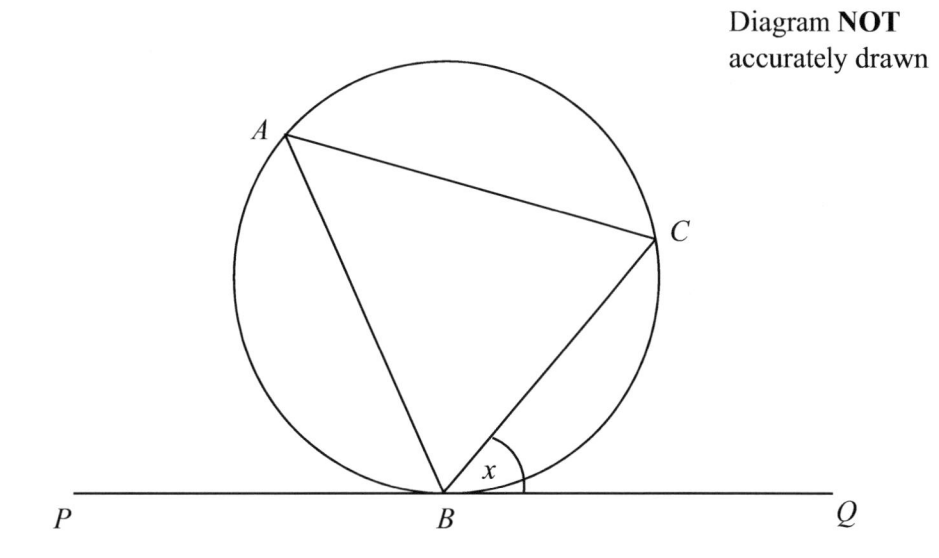

Diagram **NOT** accurately drawn

Angle *CBQ* = *x*

Prove that *AC* = *BC*

(3)

(Total for Question 14 is 3 marks)

15

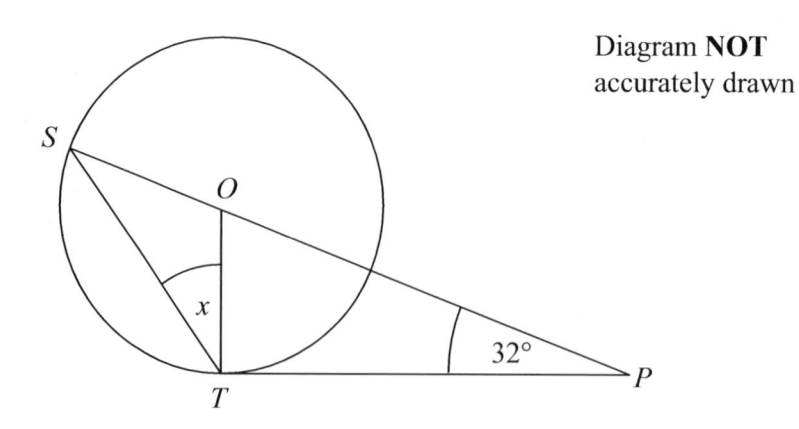

Diagram NOT accurately drawn

S and T are points on the circumference of a circle, centre O.
PT is a tangent to the circle.
SOP is a straight line.
Angle OPT = 32°

Work out the size of the angle marked x.
You must give a reason for each stage of your working.

x = ...

(Total for Question 15 is 4 marks)

Write your name here

Surname

Other names

In the style of:
Pearson Edexcel
Level 1/Level 2 GCSE (9 - 1)

Centre Number

Candidate Number

Mathematics

Cumulative Frequency

Higher Tier

GCSE style questions arranged by topic

Paper Reference
1MA1/3H

You must have: Ruler graduated in centimetres and millimetres, protractor, pair of compasses, pen, HB pencil, eraser, calculator.

Total Marks

Instructions

- Use **black** ink or ball-point pen.
- **Fill in the boxes** at the top of this page with your name, centre number and candidate number.
- Answer **all** questions.
- Answer the questions in the spaces provided
  – *there may be more space than you need.*
- **Calculators may be used.**
- If your calculator does not have a π button, take the value of π to be 3.142 unless the question instructs otherwise.
- Diagrams are **NOT** accurately drawn, unless otherwise indicated.
- You must **show all your working out**.

Information

- The total mark for this paper is
- The marks for **each** question are shown in brackets
  – *use this as a guide as to how much time to spend on each question.*

Advice

- Read each question carefully before you start to answer it.
- Keep an eye on the time.
- Try to answer every question.
- Check your answers if you have time at the end.

Turn over

© Peter Bland

1 Two groups of people are trying to lose weight.

(a) Group A join a gym.
The graph shows information about their weight loss after one month.

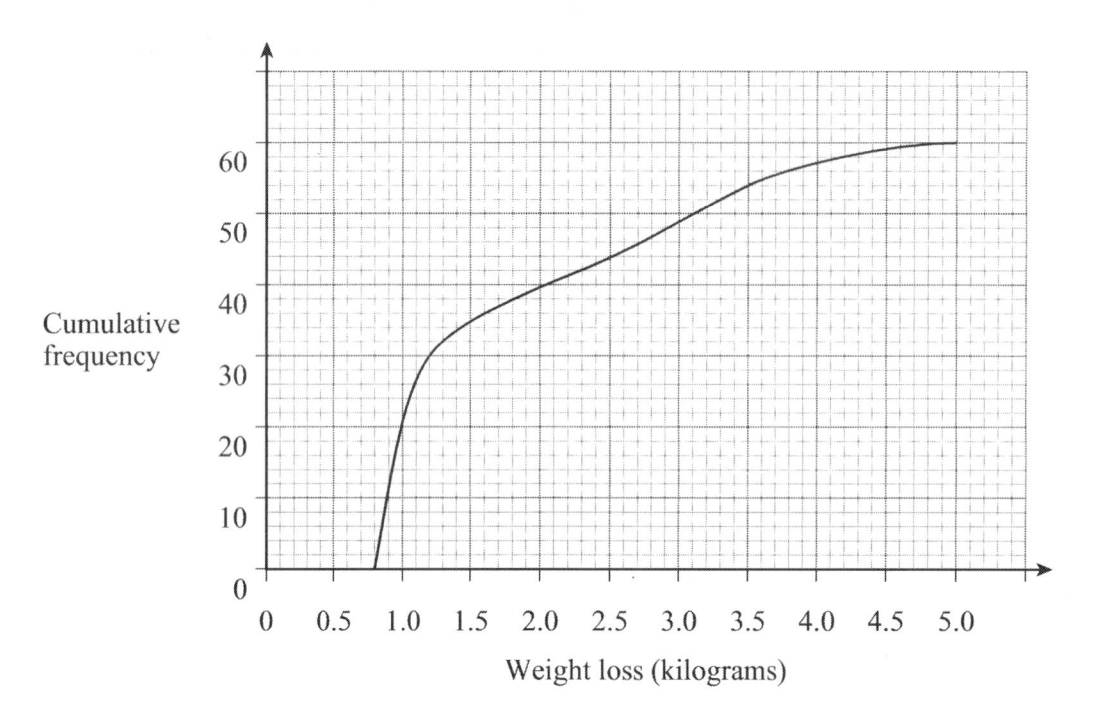

Cumulative frequency vs Weight loss (kilograms)

(i) How many people are in group A?

60 people

(1)

(ii) Does everyone in group A lose weight?
Write down how you decide.

Yes

(1)

(b) Group B follow a diet.
The box plot shows information about their weight loss after one month.

[Box plot annotated: smallest value, lower quartile, median, upper quartile, largest value. X-axis: Weight loss (kilograms), from −0.5 to 5.0]

Does everyone in group B lose weight? Write down how you decide.

No, because it goes to a negative value

(1)

(c) Compare the weight loss of group A with group B.

Compare: medians
interquartile range
* word 'spread'

(5)

Total for Question 1 is 8 marks

2. The box plot gives information about the distribution of the weights of bags on a plane.

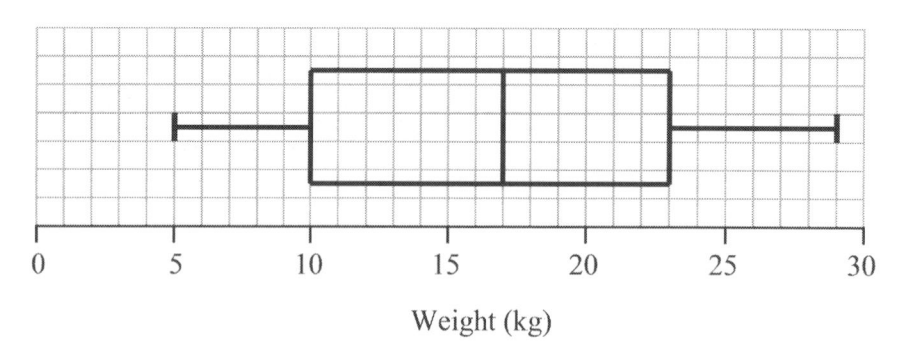

Weight (kg)

(a) Georgina says the lightest bag weighs 10 kg.

She is **wrong**.
Explain why.

The smallest value as shown in the box plot is 5kg, 10kg is the lower quartile amount.

(1)

(b) Write down the median weight.

................17................ kg
(1)

(c) Work out the interquartile range of the weights.

................10kg, 2................ kg
(1)

There are 240 bags on the plane.

(d) Work out the number of bags with a weight of 23 kg or more.

....
(2)

Total for Question 2 is 5 marks

3) David measured the height, in cm, of each tomato plant in his greenhouse.
He used the results to draw the box plot shown below.

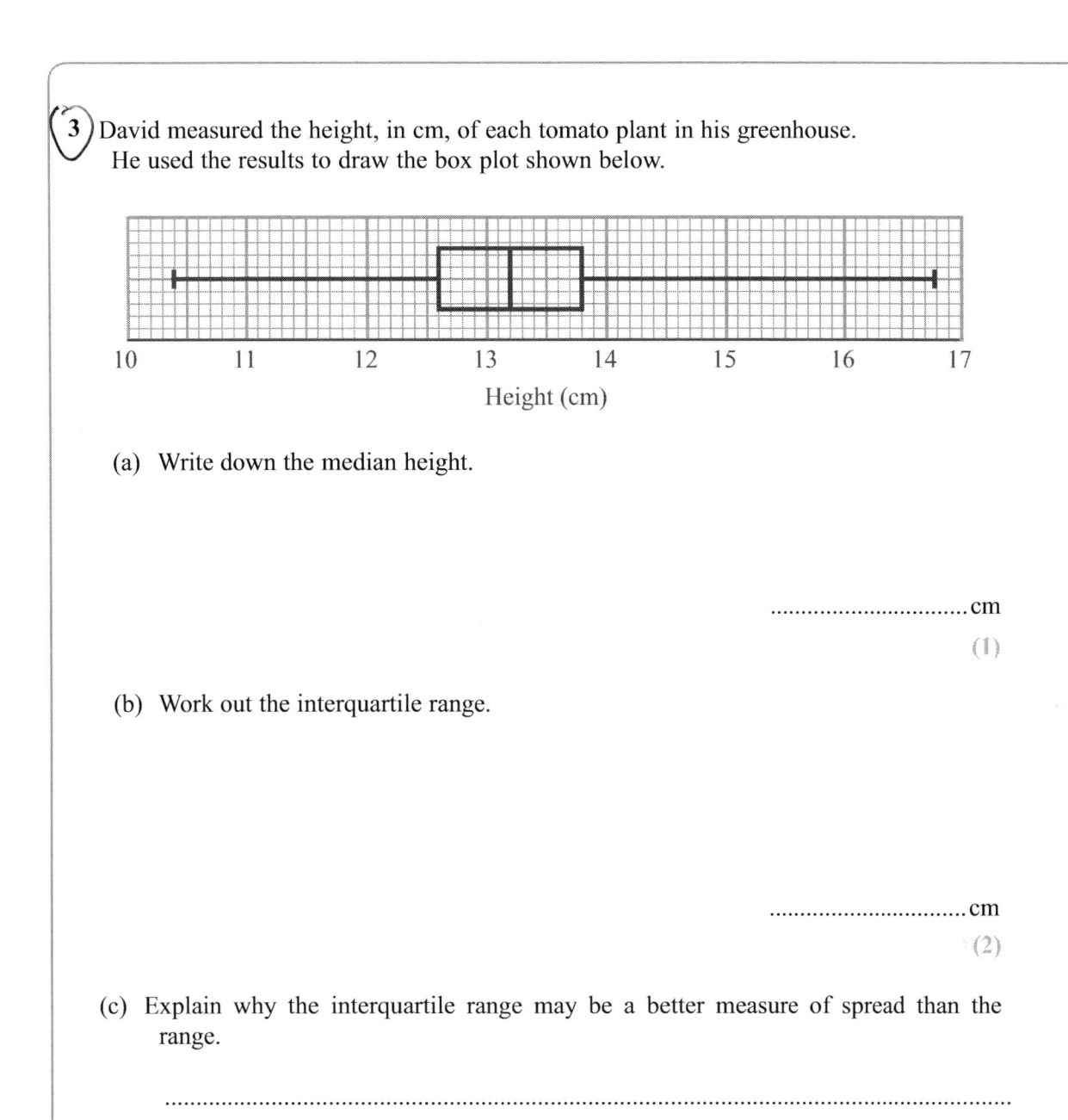

(a) Write down the median height.

..............................cm

(1)

(b) Work out the interquartile range.

..............................cm

(2)

(c) Explain why the interquartile range may be a better measure of spread than the range.

....

....

(1)

Total for Question 3 is 4 marks

4 The incomplete box plot and table show some information about some marks.

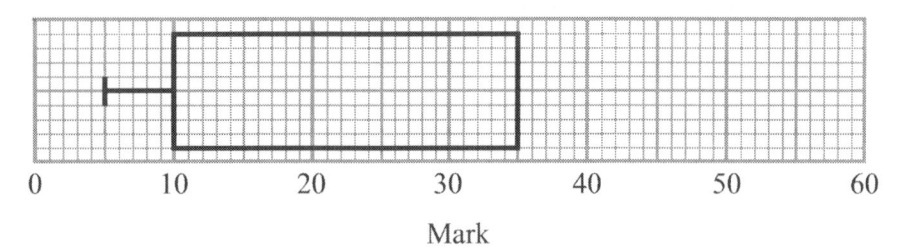

	Mark
Lowest mark	5
Lower quartile	
Median	30
Upper quartile	35
Highest mark	55

(a) Use the information in the table to complete the box plot.

(2)

(b) Use the information in the box plot to complete the table.

(1)

Total for Question 4 is 3 marks

5 The table shows a summary of the marks scored by 120 people in a test.

Mark	Frequency	Cumulative f
0 < mark ≤ 20	8	8
20 < mark ≤ 40	12	20
40 < mark ≤ 60	46	66
60 < mark ≤ 80	35	101
80 < mark ≤ 100	19	120

(a) Three-quarters of the people pass the test.

Use a cumulative frequency graph to estimate the pass mark.

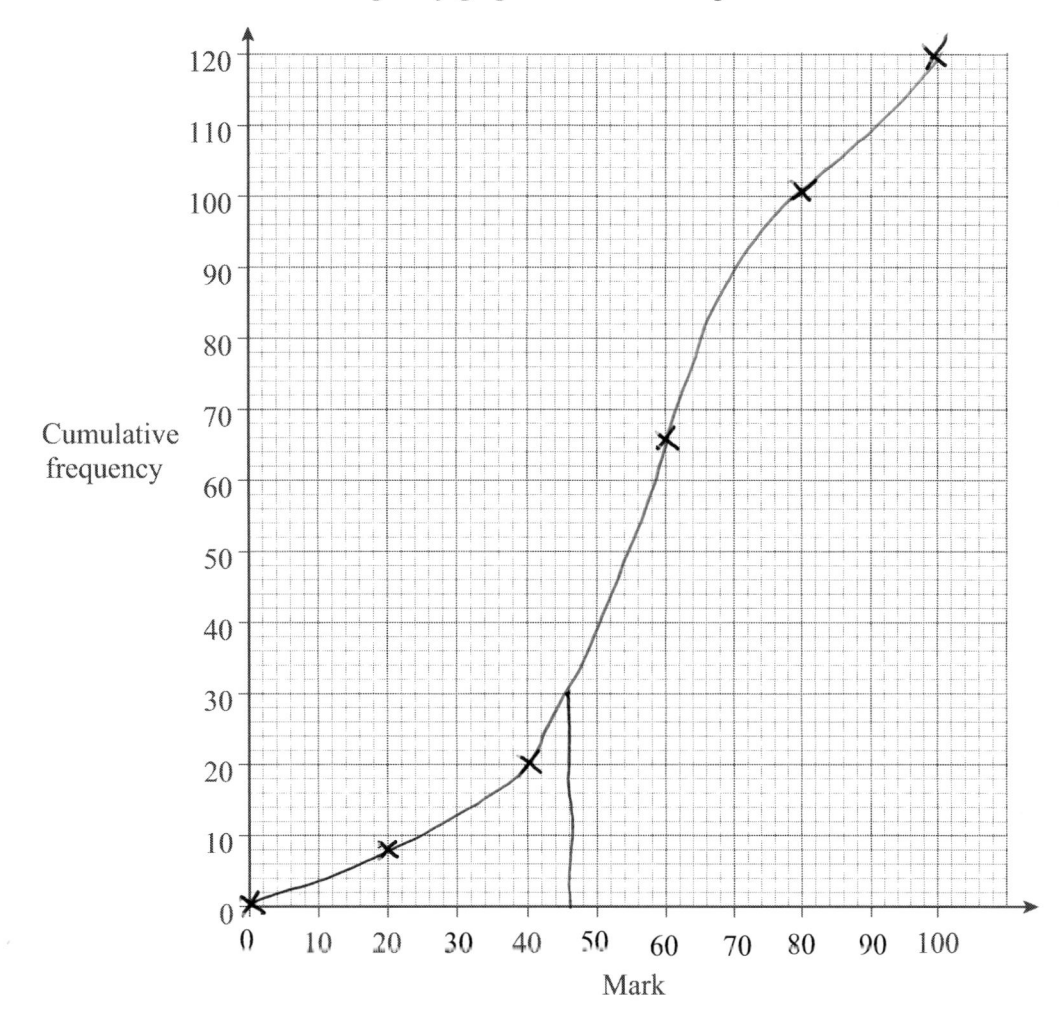

..........46..........

(5)

(b) Here is the table again.

Mark	Frequency
$0 < \text{mark} \leq 20$	8
$20 < \text{mark} \leq 40$	12
$40 < \text{mark} \leq 60$	46
$60 < \text{mark} \leq 80$	35
$80 < \text{mark} \leq 100$	19

Two of these 120 people are chosen at random.

(i) Work out the probability that both scored **over** 60.

....
(2)

(ii) Work out the probability that one scored **over** 80 and the other scored 80 or **under**.

....
(3)

Total for Question 5 is 10 marks

6 Georgina did a survey about the amounts of money spent by 120 families during summer holidays.

The cumulative frequency table gives some information about the amounts of money spent by the 120 families.

Amount (£A) spent	Cumulative frequency
$0 \leqslant A < 100$	13
$0 \leqslant A < 150$	25
$0 \leqslant A < 200$	42
$0 \leqslant A < 250$	64
$0 \leqslant A < 300$	93
$0 \leqslant A < 350$	110
$0 \leqslant A < 400$	120

(a) On the grid, draw a cumulative frequency diagram.

(2)

(b) Use your cumulative frequency diagram to estimate the median.

£ *250*

(2)

A survey of the amounts of money spent by 200 families during their Christmas holidays gave a median of £305

(c) Compare the amounts of money spent at Christmas with the amounts of money spent in summer.

They spend more money at Christmas

(1)

Cumulative frequency

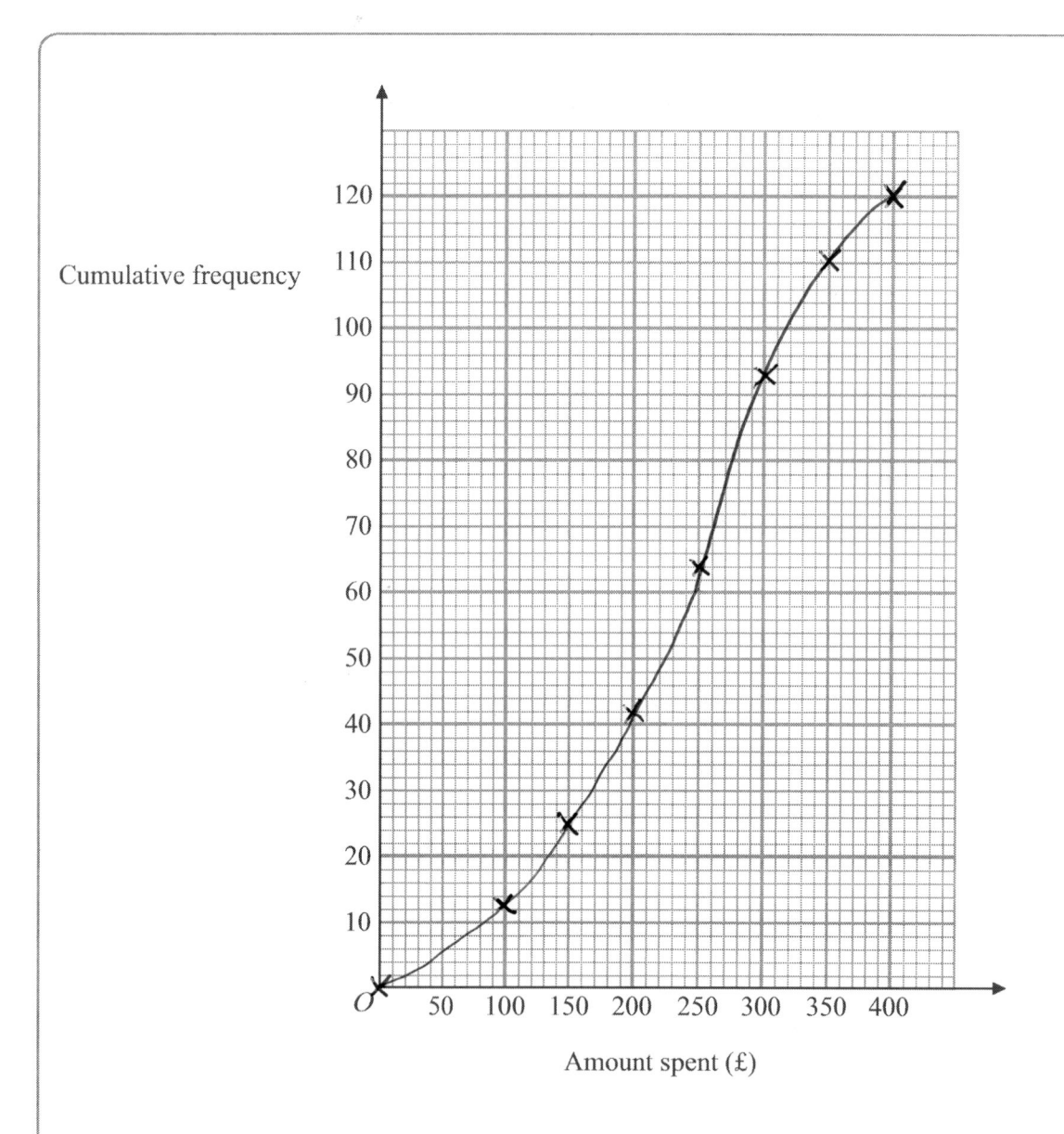

Amount spent (£)

Total for Question 6 is 5 marks

7 The table shows information about the number of felt tip pens in 100 childrens pencil cases.

Number of pens	Frequency
$0 < n \leq 20$	18
$20 < n \leq 40$	22
$40 < n \leq 60$	35
$60 < n \leq 80$	15
$80 < n \leq 100$	8
$100 < n \leq 120$	2

(a) Complete the cumulative frequency table for this information.

Number of pens	Cumulative frequency
$0 < n \leq 20$	18
$0 < n \leq 40$	
$0 < n \leq 60$	
$0 < n \leq 80$	
$0 < n \leq 100$	
$0 < n \leq 120$	

(1)

(b) On the grid, draw a cumulative frequency graph for your table.

(2)

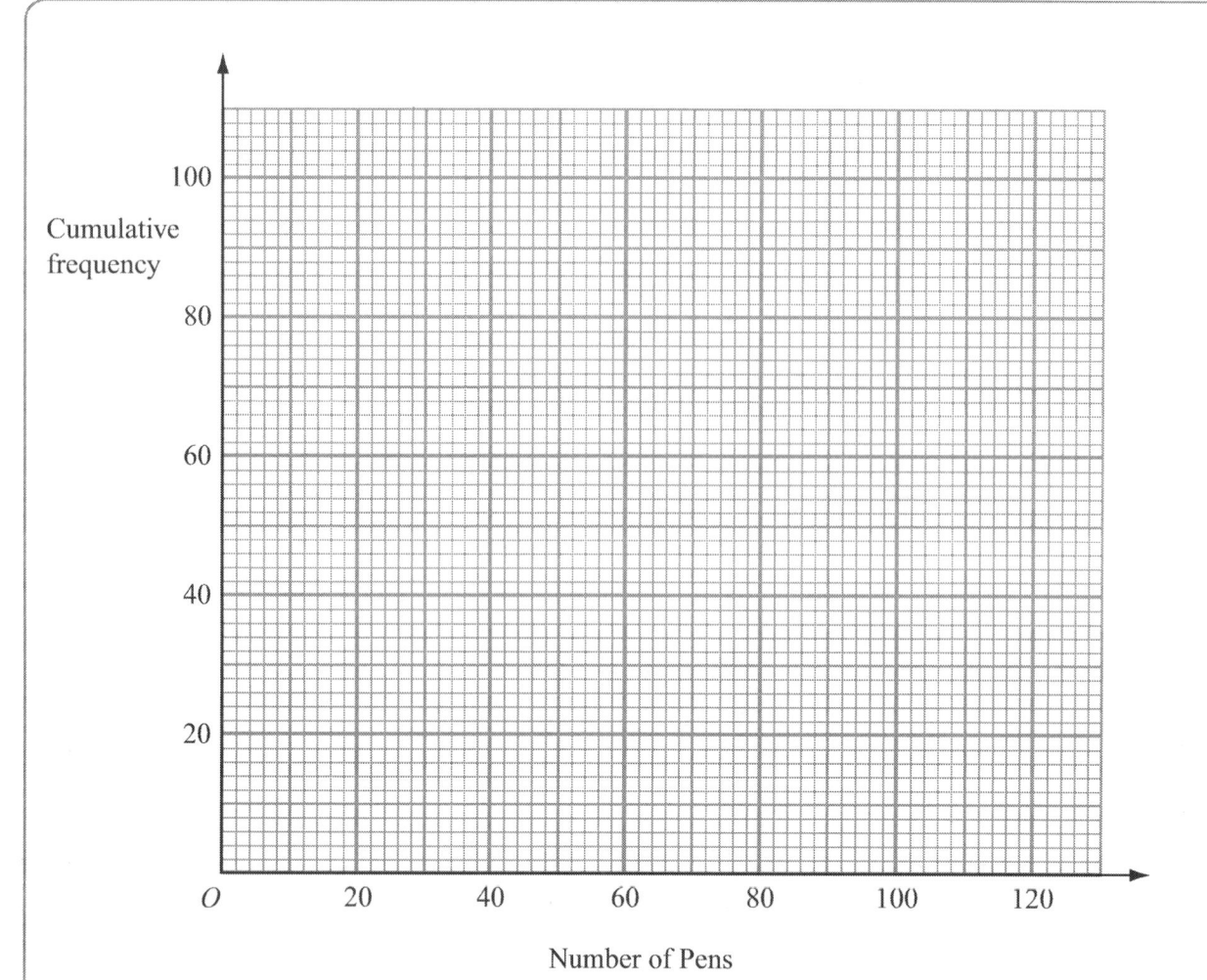

(c) Use your graph to find an estimate for the median number of pens.

....
(1)

Total for Question 7 is 4 marks

8 A company tested 100 batteries.

The table shows information about the number of hours that the batteries lasted.

Time (t hours)	Frequency
$50 \leqslant t < 55$	12
$55 \leqslant t < 60$	21
$60 \leqslant t < 65$	36
$65 \leqslant t < 70$	23
$70 \leqslant t < 75$	8

(a) Complete the cumulative frequency table for this information.

(1)

Time (t hours)	Cumulative frequency
$50 \leqslant t < 55$	12
$50 \leqslant t < 60$	
$50 \leqslant t < 65$	
$50 \leqslant t < 70$	
$50 \leqslant t < 75$	

(b) On the grid, draw a cumulative frequency graph for your completed table.

(2)

(c) Use your completed graph to find an estimate for the median time.
You must state the units of your answer.

....
(2)

Cumulative frequency

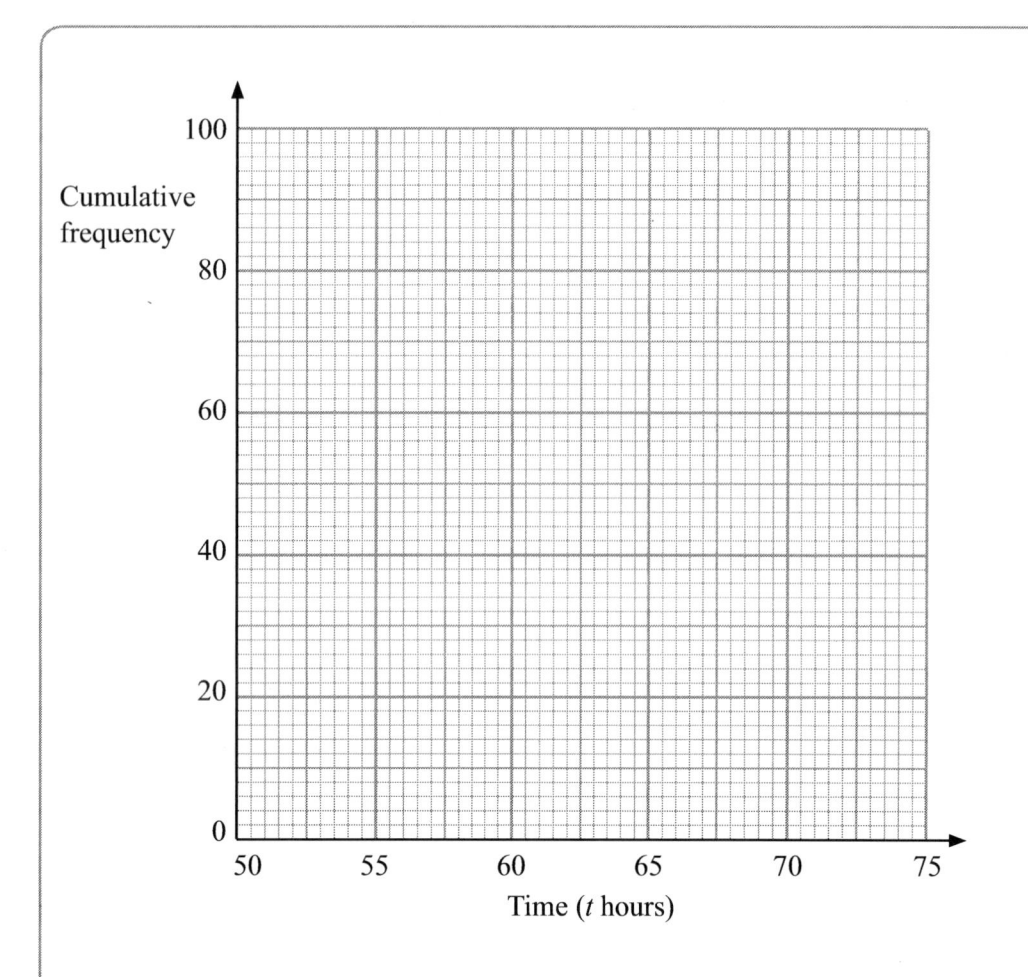

(blank grid with x-axis "Time (*t* hours)" from 50 to 75, y-axis from 0 to 100)

Total for Question 8 is 5 marks

9 The table gives some information about the number of fish caught in a match.

Number of fish	Frequency
$0 < n \leq 20$	16
$20 < n \leq 30$	26
$30 < n \leq 40$	23
$40 < n \leq 50$	10
$50 < n \leq 60$	5

(a) Write down the modal class interval.

....
(1)

(b) Complete the cumulative frequency table.

Number of fish	Cumulative Frequency
$0 < n \leq 20$	
$0 < n \leq 30$	
$0 < n \leq 40$	
$0 < n \leq 50$	
$0 < n \leq 60$	

(1)

(c) On the grid opposite, draw a cumulative frequency graph for your table.

(2)

(d) Use your graph to find an estimate for

(i) the median number of fish,

.... minutes

(ii) the interquartile range of the number of fish.

.... minutes
(3)

79

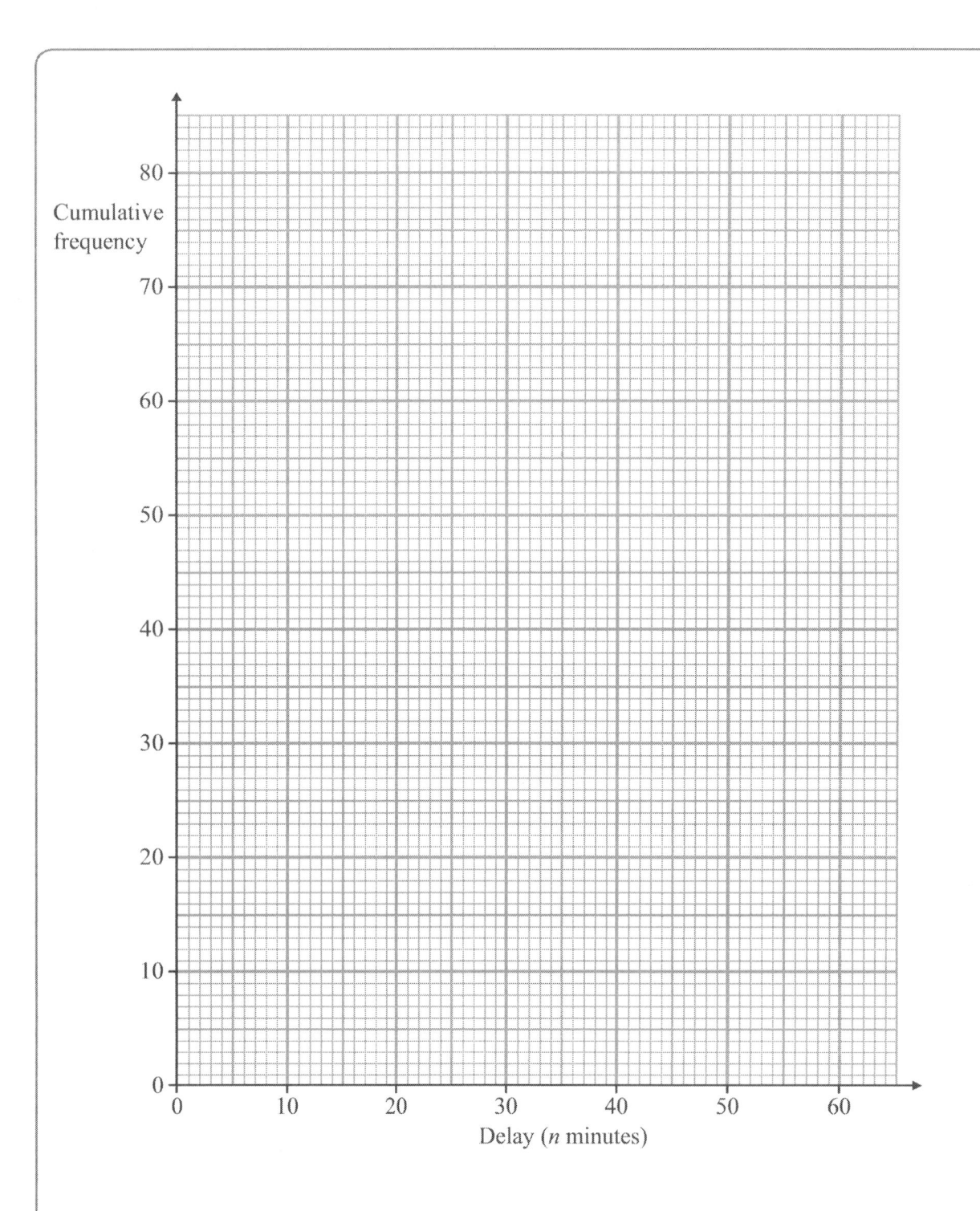

Total for Question 9 is 7 marks

10 Here are four cumulative frequency diagrams.

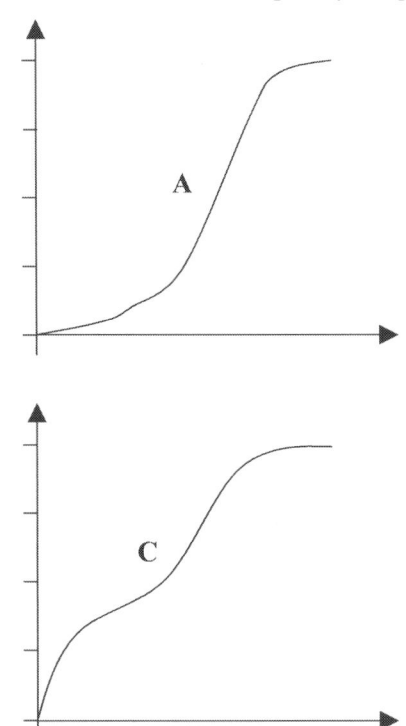

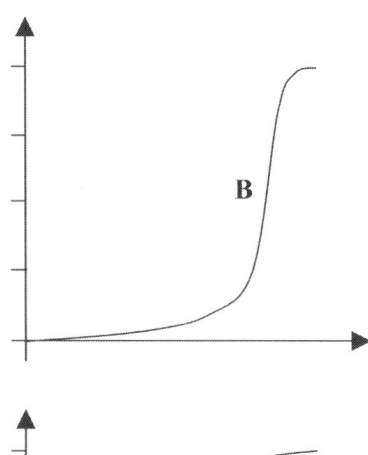

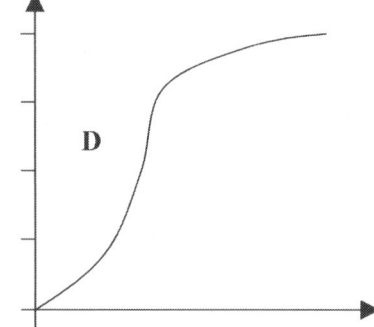

Here are four box plots.

For each box plot, write down the letter of the appropriate cumulative frequency diagram.

P and ....

Q and ....

R and ....

S and ....

Total for Question 10 is 2 marks

11 The table shows information about the time, *m* millimetres 120 tomato plants grow in a week.

Time (*m* millimetres)	Frequency
$70 < m \leqslant 80$	4
$80 < m \leqslant 90$	12
$90 < m \leqslant 100$	34
$100 < m \leqslant 110$	32
$110 < m \leqslant 120$	26
$120 < m \leqslant 130$	12

(a) Write down the modal class interval.

....
(1)

(b) Complete the cumulative frequency table.

Time (*m* millimetres)	Cumulative frequency
$70 < m \leqslant 80$	4
$70 < m \leqslant 90$	
$70 < m \leqslant 100$	
$70 < m \leqslant 110$	
$70 < m \leqslant 120$	
$70 < m \leqslant 130$	

(1)

(c) On the grid, draw a cumulative frequency graph for your cumulative frequency table.

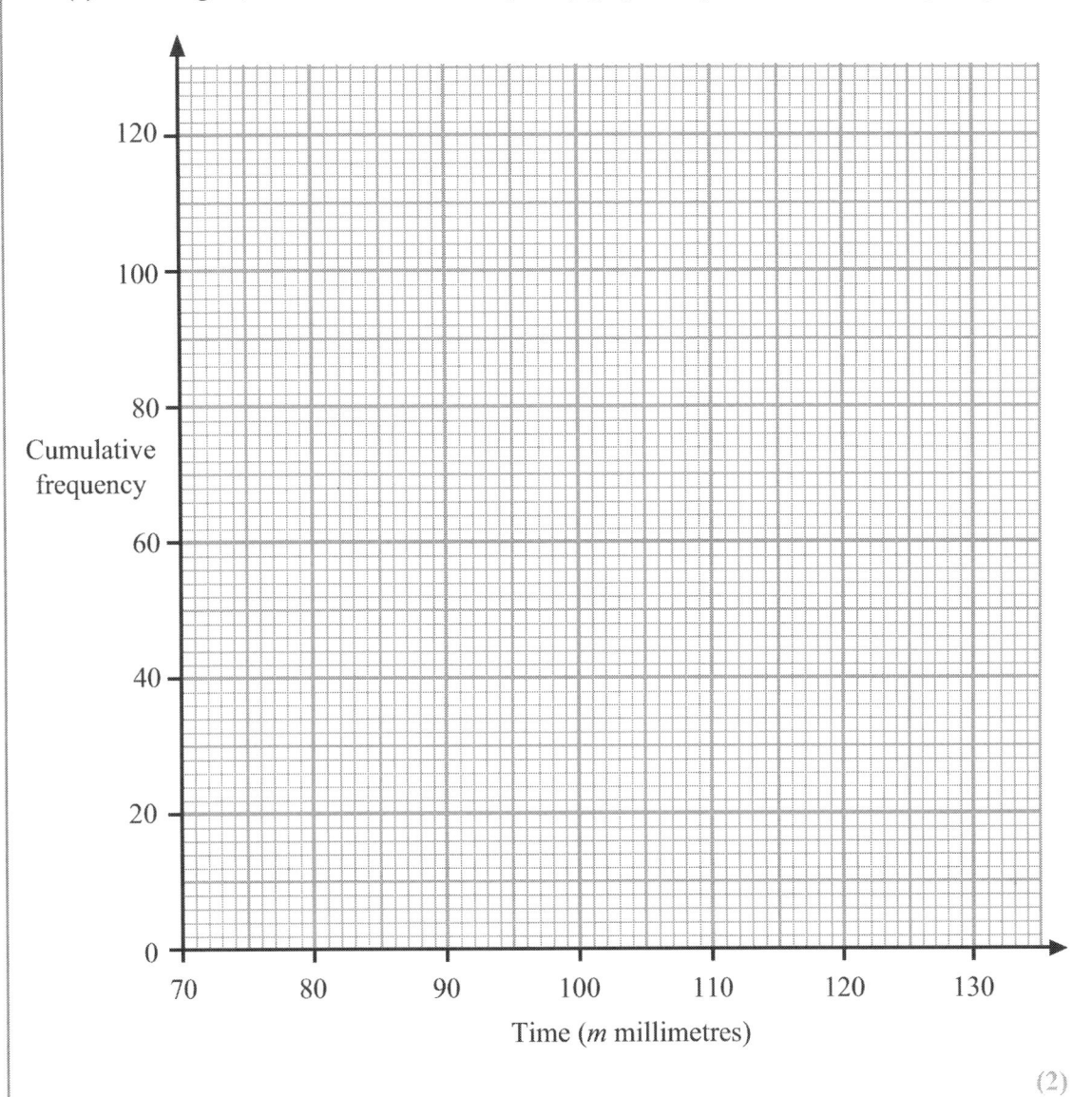

(2)

(d) Use your graph to find an estimate for the median.

.... minutes

(1)

Total for Question 11 is 5 marks

12 The cumulative frequency table shows the marks some students got in a test.

Mark (m)	Cumulative frequency
$0 < m \leqslant 10$	8
$0 < m \leqslant 20$	23
$0 < m \leqslant 30$	48
$0 < m \leqslant 40$	65
$0 < m \leqslant 50$	74
$0 < m \leqslant 60$	80

(a) On the grid, plot a cumulative frequency graph for this information.

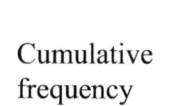

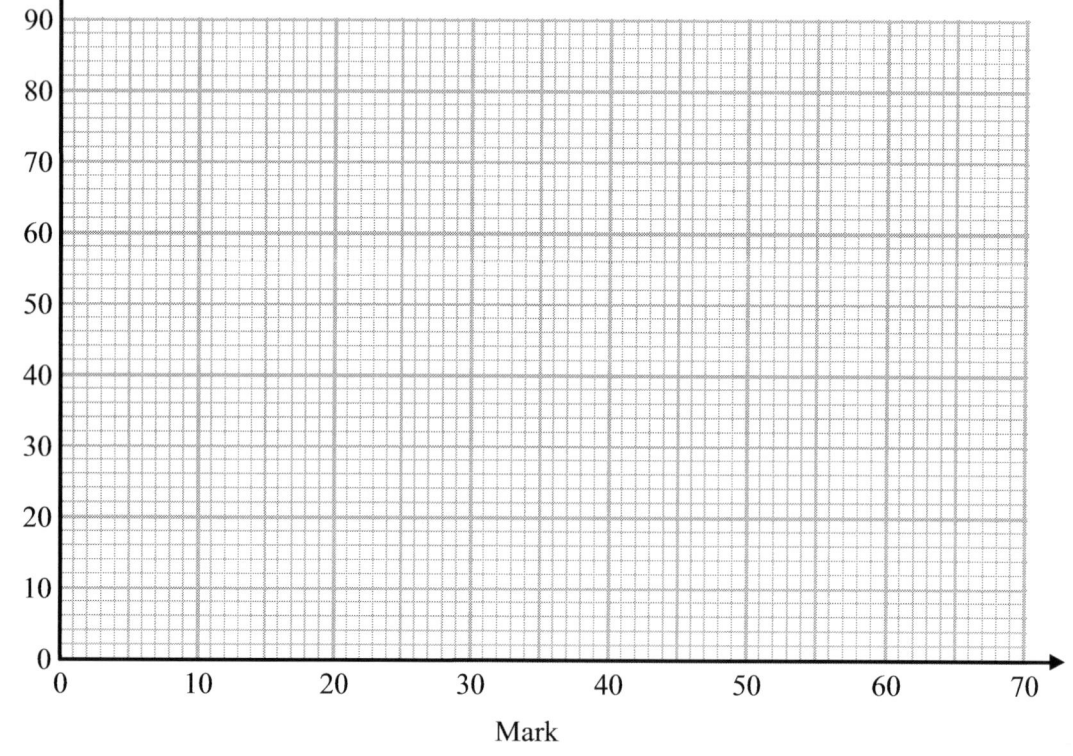

(2)

(b) Find the median mark.

....

(1)

Students either pass the test or fail the test.
The pass mark is set so that 3 times as many students fail the test as pass the test.

(c) Find an estimate for the lowest possible pass mark.

....
(3)

(Total for Question 12 is 6 marks)

13 David measures the heights of 80 plants he has grown.
This table summarises his results.

Height, h cm	$0 < h \leq 50$	$50 < h \leq 100$	$100 < h \leq 125$	$125 < h \leq 150$
Number of plants	8	38	31	3

(a) (i) Complete the cumulative frequency table below.

Height, h cm	$h \leq 50$	$h \leq 100$	$h \leq 125$	$h \leq 150$
Cumulative frequency	8			

(2)

(ii) Draw the cumulative frequency graph.

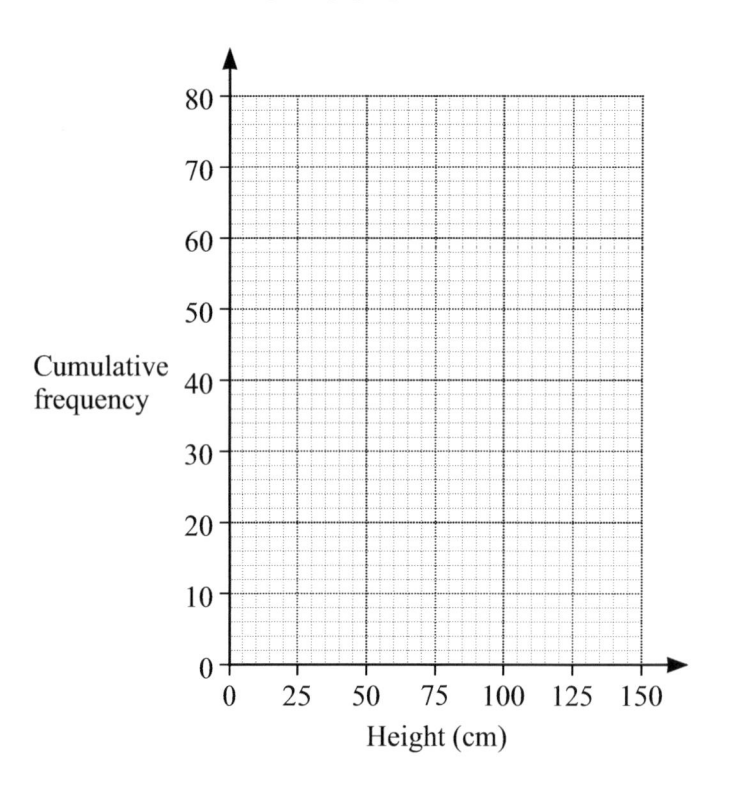

(2)

(b) Tara asks if David has 10 plants over 120 cm in height.

Explain why David cannot be certain that he has 10 plants over this height.

....

....

(1)

(c) David sells these 80 plants using the price list below.

Height, h cm	$h \leq 80$	$80 < h \leq 120$	$h > 120$
Price (£)	2.00	3.50	5.00

Each plant costs him 60p to grow.

Estimate the total profit David will receive when he sells all these plants.

£ ....

(6)

Total for Question 13 is 11 marks

14 The table shows the marks gained by 150 students taking an examination.

Mark (m)	0<m≤10	10<m≤20	20<m≤30	30<m≤40	40<m≤50	50<m≤60	60<m≤70	70<m≤80
Frequency	9	14	26	27	25	22	17	10

(a) (i) Construct a cumulative frequency table.

Mark (m)	m ≤ 10	m ≤ 20	m ≤ 30	m ≤ 40	m ≤ 50	m ≤ 60	m ≤ 70	m ≤ 80
Cumulative Frequency	9							150

(ii) Draw the cumulative frequency graph on the grid below.

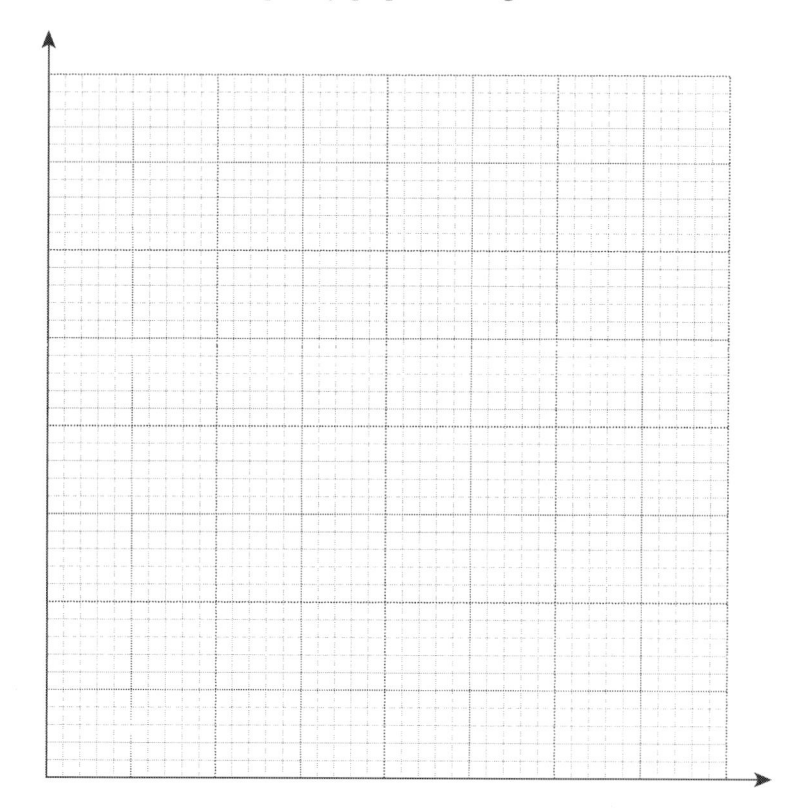

(2)

(4)

(b) Students are to be awarded Gold, Silver, Bronze or Fail.
The students' teacher wishes to award the top 10% of students Gold, the next 60% Silver and the next 20% Bronze.

Use your graph to estimate the lowest mark that Silver will be awarded for.

(b) ....
(3)

(c) Explain why the teacher's method will not necessarily award Gold to exactly 10% of the students.

....

....
(1)

Total for Question 14 is 10 marks

Write your name here

Surname

Other names

In the style of:
Pearson Edexcel
Level 1/Level 2 GCSE (9 - 1)

Centre Number

Candidate Number

Mathematics

Fractions

Foundation Tier

GCSE style questions arranged by topic

Paper Reference
1MA1/2F

You must have: Ruler graduated in centimetres and millimetres, protractor, pair of compasses, pen, HB pencil, eraser, calculator.

Total Marks

Instructions

- Use **black** ink or ball-point pen.
- **Fill in the boxes** at the top of this page with your name, centre number and candidate number.
- Answer **all** questions.
- Answer the questions in the spaces provided
  – there may be more space than you need.
- **Calculators may be used.**
- If your calculator does not have a π button, take the value of π to be 3.142 unless the question instructs otherwise.
- Diagrams are **NOT** accurately drawn, unless otherwise indicated.
- You must **show all your working out.**

Information

- The total mark for this paper is
- The marks for **each** question are shown in brackets
  – use this as a guide as to how much time to spend on each question.

Advice

- Read each question carefully before you start to answer it.
- Keep an eye on the time.
- Try to answer every question.
- Check your answers if you have time at the end.

Turn over

© Peter Bland

1. David earns a salary of £3500 per month.

   He gets a pay rise of 4%.

   Work out his new monthly salary.

   10% = 350
   1% = 35
   4% = 140

   3500 + 140 = 3640

   ......£3640......

   (3)

   (Total for Question 1 is 3 marks)

2(a)

    Helen wins a race.

    Her time is recorded as 50.36 seconds.

    Andrew comes second in the race.

    His time is three-hundredths of a second slower.

    Work out Andrew's time.

(2)

(b)   Round Helen's time of 50.36 seconds to 1 decimal place.

(1)

(Total for Question 2 is 6 marks)

3 Write a number in each box to make correct statements.

(a) $50\% = \dfrac{\square}{2}$

(1)

(b) $0.3 = \dfrac{\square}{10}$

(1)

(c) $\dfrac{1}{3} = \dfrac{\square}{9}$

(1)

(d) $\dfrac{3}{15} = \dfrac{\square}{5}$

(1)

(Total for Question 3 is 4 marks)

4   Two banks calculate the yearly interest they pay customers.

Westminster Bank

4% of the total that you invest

For example: Invest £700

Interest = 4% of £700

District Bank

1% of the first £300 that you invest 6% of amounts over £300 that you invest

For example: Invest £700

Interest = 1% of £300 + 6% of £400

Ashna has £500 to invest for one year.

Work out which bank will pay her more interest. State how much **extra** interest she will earn.

Bank ....

Extra Interest £................................

(Total for Question 4 is 5 marks)

5   There are 180 people at a wedding. 20% are children. One-half are men. The rest are women.

How many women are at the wedding?

....
(4)

(Total for Question 5 is 4 marks)

6 (a) Shade $\frac{9}{25}$ of this square grid.

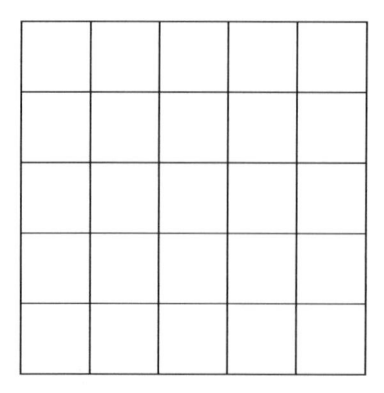

(1)

(b) Shade $\frac{4}{5}$ of this square grid.

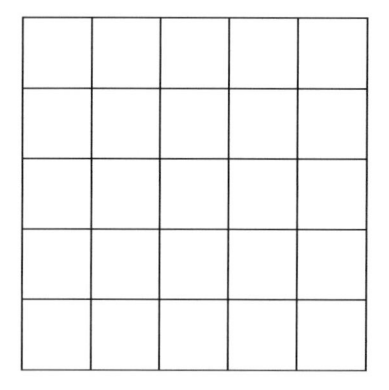

(1)

(c) Use your answers to part (a) and part (b) to write down the answer to $\frac{4}{5} - \frac{9}{25}$

....

(1)

(d) Work out $\frac{2}{3}$ of 36

....

(2)

(Total for Question 6 is 5 marks)

7 (a) Use your calculator to work out $\dfrac{4.7}{9.4-3.5}$

Write down all the figures on your calculator display.

....
(2)

(b) Write these numbers in order of size.
Start with the smallest number.

$$0.82 \qquad \frac{4}{5} \qquad 85\% \qquad \frac{2}{3} \qquad \frac{7}{8}$$

....
(2)

(Total for Question 7 is 4 marks)

8 A concert ticket costs £65 plus a booking charge of 15%.

Work out the total cost of a concert ticket.

£ ....

(Total for Question 8 is 3 marks)

9 A school canteen sells salads and hot meals.
In one week the number of salads sold and the number of hot meals sold were in the ratio 3 : 5

The total number of salads and hot meals sold in the week was 1456

Work out the number of salads sold.

....

(Total for Question 9 is 2 marks)

10 A garage sells British cars and foreign cars.
The ratio of the number of British cars sold to the number of foreign cars sold is 2 : 7

The garage sells 45 cars in one week.

(a) Work out the number of British cars the garage sold that week.

....
(2)

A car tyre costs £80 plus VAT at $17\frac{1}{2}$ %.

(b) Work out the total cost of the tyre.

£ ....
(3)

The value of a new car is £14 000

The value of the car depreciates by 20% per year.

(c) Work out the value of the car after 2 years.

£ ....
(3)

(Total for Question 10 is 8 marks)

11 There are some pens in a bag.

36 of the pens are blue.

24 of the pens are black.

(a) Write down the ratio of the number of blue pens to the number of black pens.
Give your ratio in its simplest form.

.... : ....
(2)

There are some books and comics in a box.
The total number of books and comics is 54
The ratio of the number of books to the number of comics is 1 : 5

(b) Work out the number of books in the box.

....
(2)

(Total for Question 11 is 4 marks)

12 Louis invested £6500 for 2 years in a savings account.

He was paid 4% per annum compound interest.

(a) How much did Louis have in his savings account after 2 years?

£ ....
(3)

Hassan invested £2400 for n years in a savings account.

He was paid 7.5% per annum compound interest.

At the end of the n years he had £3445.51 in the savings account.

(b) Work out the value of n.

....
(2)

(Total for Question 12 is 5 marks)

13 Work out $\dfrac{2}{3} \div \dfrac{7}{9}$

Give your fraction in its simplest form.

....
(3)

(b) Work out $2\dfrac{1}{3} - 1\dfrac{2}{5}$

....
(3)

(Total for Question 13 is 6 marks)

14 (a) Write $\frac{7}{16}$ as a decimal.

....

(Total for Question 14 is 1 mark)

15 *ABCD* is a square.
This diagram is drawn accurately.

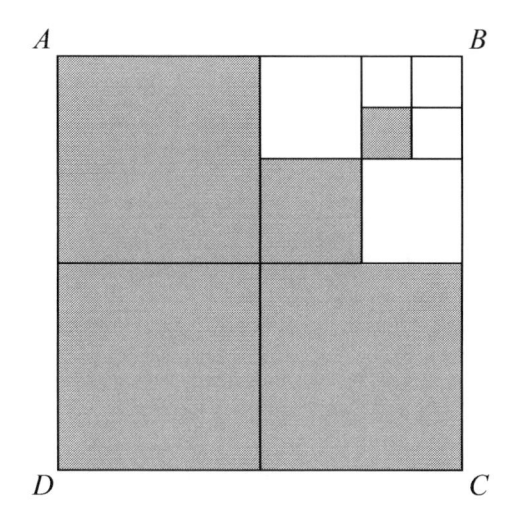

What fraction of the square *ABCD* is shaded?

....

(Total for Question 15 is 2 marks)

16 Write the following numbers in order of size.
Start with the smallest number.

0.61   0.1   0.16   0.106

....

(Total for Question 16 is 1 mark)

17 Write 0.037 as a fraction.

....

(Total for Question 17 is 1 mark)

18 Work out 15% of 80

....

(Total for Question 18 is 2 marks)

19 (a) Work out $\frac{2}{7} + \frac{1}{5}$

....
(2)

(b) Work out $1\frac{2}{3} \div \frac{3}{4}$

....
(2)

(Total for Question 19 is 4 marks)

20 In a sale, normal prices are reduced by 20%.
The normal price of a coat is reduced by £15

Work out the normal price of the coat.

£.................................

(Total for Question 20 is 2 marks)

21 Work out 6.34×5.2

(Total for Question 21 is 3 marks)

22 Write the following numbers in order of size, smallest first.

60.6   6.601   6.106   0.6   6.06

....
smallest

(Total for Question 22 is 3 marks)

23 Nikki organised a wedding.
Guests had to choose their meal from pasta, chicken or beef.

- $\frac{1}{3}$ of the guests chose pasta.
- $\frac{5}{12}$ of the guests chose chicken.
- 24 of the guests chose beef.

How many guests were at the wedding?

....

(Total for Question 23 is 4 marks)

24 Sophie took a maths test. She scored 28 marks out of 40.

Tara took an English test. She scored 32 marks out of 47.

Tara said

> I did better than Sophie as I scored more marks.

By writing each score as a percentage, show that Tara is wrong.

(Total for Question 24 is 3 marks)

25 A shop has a sale that offers 20% off all prices.

On the final day they reduce all sale prices by 25%.

Esta buys a hairdryer on the final day.
Work out the **overall** percentage reduction on the price of the hairdryer.

.... %

(Total for Question 25 is 6 marks)

26   Write these in order, smallest first.

$$0.34 \qquad \frac{1}{3} \qquad 3.5\%$$

....
smallest

(Total for Question 26 is 2 marks)

27   Carol drinks $\frac{3}{8}$ of a litre of milk each breakfast.

Milk costs 89p for a 2-litre carton and 49p for a 1-litre carton.

What is the smallest amount that Carol would have to spend to buy milk for one week?
Show your working.

£ ....

(Total for Question 27 is 3 marks)

28 How many centimetres are there in 3.7 metres?

Circle your answer.

0.037   37   370

(1)

(Total for Question 28 is 1 mark)

29 Circle the fraction that is not equivalent to $\frac{3}{8}$

$\frac{6}{16}$ $\frac{9}{24}$ $\frac{12}{32}$ $\frac{15}{35}$

(1)

Here are some numbers.

9.6   12.6   15.4   7.6   12.4   17.4

Write the numbers in pairs so that the sum of the numbers in each pair is the same.

.... and ....

.... and ....

.... and ....

(Total for Question 29 is 3 marks)

30 Work out 51% of 400

....
(Total for Question 30 is 2 marks)

31 Write 180 g as a fraction of 3 kg.

Give your answer in its simplest form.

....
(Total for Question 31 is 2 marks)

32 Work out $2\frac{3}{4} \times 1\frac{5}{7}$

Give your answer as a mixed number in its simplest form.

....
(Total for Question 32 is 3 marks)

33 A gym has 275 members.

   40% are bronze members.

   28% are silver members.

   The rest are gold members.

Work out the number of gold members.

....

(Total for Question 33 is 3 marks)

34 120 men and 80 women were asked if they drive to work.

   Altogether $\frac{1}{4}$ of the people said yes.

   $\frac{1}{3}$ of the men said yes.

What fraction of the women said yes?

....

(Total for Question 34 is 4 marks)

35 24 boys, 45 girls and 281 adults are the members of a tennis club. 50 more children join the club.

The number of girls is **now** 18% of the total number of members.

How many of the 50 children were **boys**?

....

(Total for Question 35 is 4 marks)

---

36 Circle the decimal that is closest in value to $\dfrac{2}{3}$

(1)

      0.6   0.66   0.667   0.67

(Total for Question 36 is 1 mark)

---

37 In 1999 the minimum wage for adults was £3.60 per hour.

In 2013 it was £6.31 per hour.

Work out the percentage increase in the minimum wage.

.... %

(Total for Question 37 is 3 marks)

Write your name here

Surname

Other names

In the style of:
Pearson Edexcel
Level 1/Level 2 GCSE (9 - 1)

Centre Number

Candidate Number

Mathematics

Frequency

Foundation Tier

GCSE style questions arranged by topic

Paper Reference
1MA1/3F

You must have: Ruler graduated in centimetres and millimetres, protractor, pair of compasses, pen, HB pencil, eraser, calculator.

Total Marks

Instructions

- Use **black** ink or ball-point pen.
- **Fill in the boxes** at the top of this page with your name, centre number and candidate number.
- Answer **all** questions.
- Answer the questions in the spaces provided
  – there may be more space than you need.
- **Calculators may be used.**
- If your calculator does not have a π button, take the value of π to be 3.142 unless the question instructs otherwise.
- Diagrams are **NOT** accurately drawn, unless otherwise indicated.
- You must **show all your working out**.

Information

- The total mark for this paper is
- The marks for **each** question are shown in brackets
  – use this as a guide as to how much time to spend on each question.

Advice

- Read each question carefully before you start to answer it.
- Keep an eye on the time.
- Try to answer every question.
- Check your answers if you have time at the end.

Turn over

© Peter Bland

1 (a) Basil records the types of fish that he caught during his holiday in The Bahamas.

   (i) Complete the table.

Type of fish	Tally	Frequency
Mutton Fish	IIII	
Grouper	III	
Jack	HHH HHH II	
Schoolmaster	HHH IIII	
	Total	

   (3)

   (ii) What fraction of the fish are Mutton Fish?
   Give your answer in its simplest form.

   ....
   (2)

(b) This table shows the types of fish that Peter caught during the holiday.

Type of fish	Mutton Fish	Grouper	Jack	Schoolmaster
Frequency	4	6	5	3

   He has finished the first row of a pictogram to show the results.

   Complete the key and pictogram.

   Key: 🐟 represents .... fish

Mutton Fish	🐟 🐟
Grouper	
Jack	
Schoolmaster	

   (4)

(c) 500 000 people record the types of birds in their gardens. In total, they record eight million birds.

On average, how many birds does each person record?

....
(3)

(d) Here is a list of the birds at a bird table.

| robin | robin | sparrow | blackbird | starling |
| blackbird | starling | blackbird | robin | blackbird |

One bird flies away. Another bird arrives at the bird table.
The new mode is robin.

What type of bird flies away and what type of bird arrives? Complete the table.

	Type of bird
Flies away	
Arrives	

(2)

(Total for Question 1 is 14 marks)

2 (a) The bar chart shows the amounts Isaac saves in May, June and August 2010.

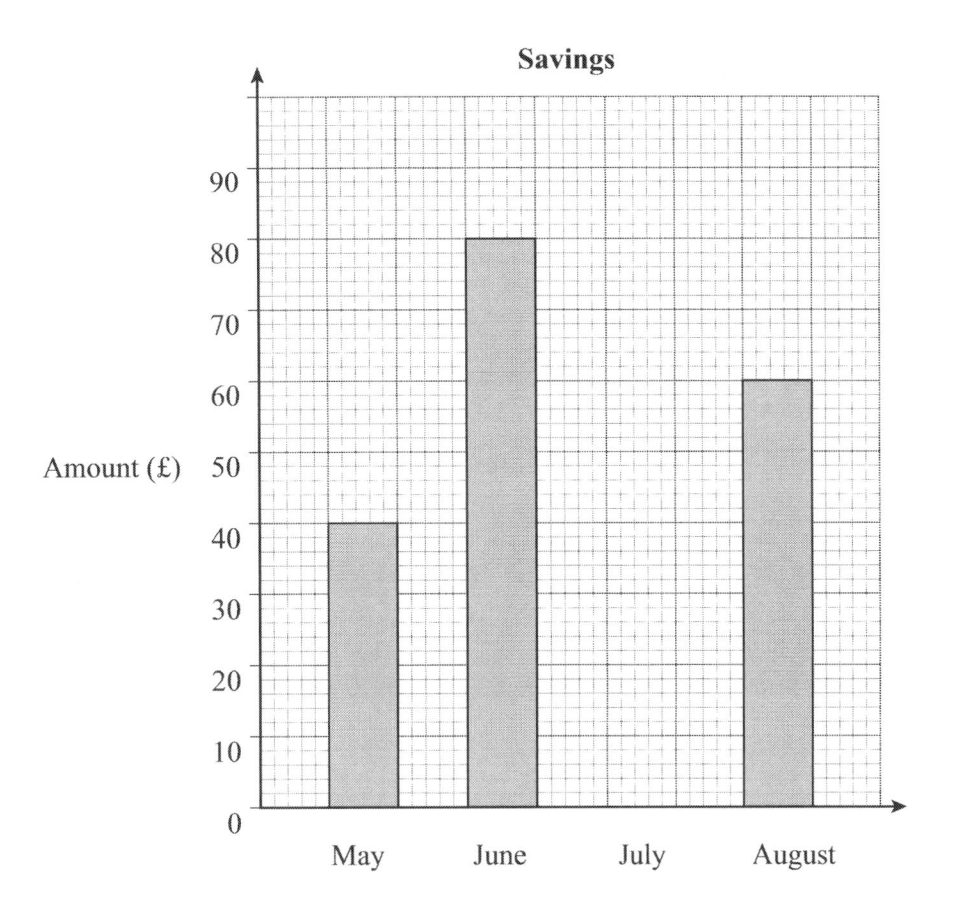

Savings

[Bar chart: Amount (£) vs months. May = 40, June = 80, August = 60]

(i) How much does he save in May 2010?

£ ....
(1)

(ii) From May to August he saves £250 in total.

Complete the bar chart by drawing the bar for July.

(3)

(b) The pictogram shows the amounts Isaac saves in the next four months.

Key: ▭▭ represents £20

September	▭▭ ▭▭ ▭▭ ▭▭
October	▭▭ ▭▭ ▭
November	▭▭ ▭
December	▭▭ ▭▭ ▭▭ ▭

Work out the range of the amount he saves in these four months. You **must** show your working.

£ ....
(2)

(c) (i) For the next 4 months he saves £50 each month.

How much has he saved in total?

£ ....
(3)

(ii) Isaac spends 50% of these total savings to pay for a holiday.

How much does he pay for the holiday?

....

....

£ ....
(2)

(Total for Question 2 is 11 marks)

3   Is money discrete or continuous? Tick a box.

☐ Discrete   ☐ Continuous

Give a reason for your answer.

(1)

Peter sells revision guides on a website. The sales in May are shown.

Sales (£)	Frequency
8	10
10	18
12	7
15	4
20	1

(a) Calculate his mean price.

£ ....
(3)

(b) Peter says that his modal price and his median price are both £10. Is he correct?
Give reasons and working to show how you decide.

(2)

(c) Georgina also sells revision guides on a website

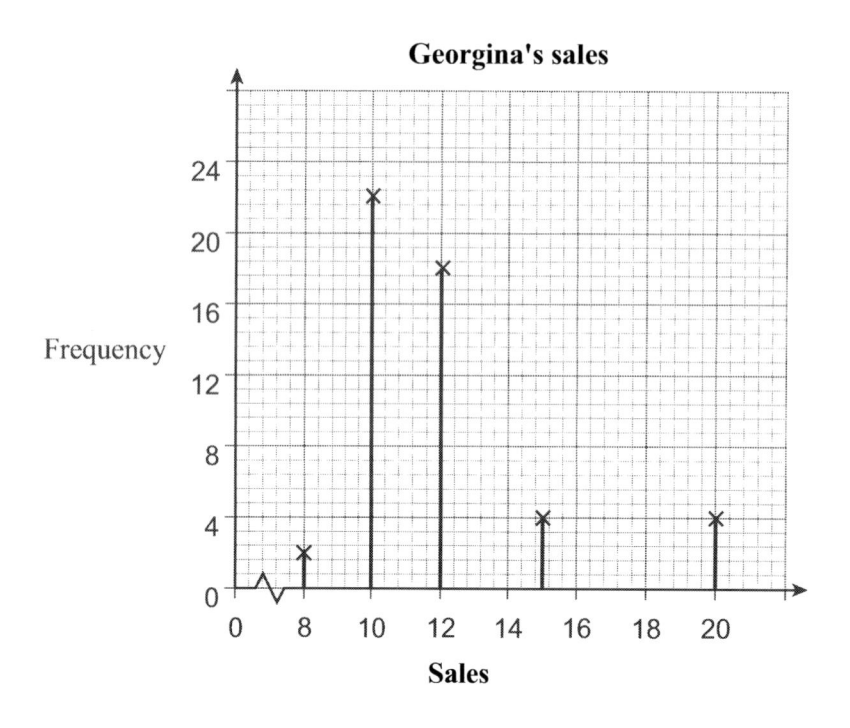

Georgina's sales

[Vertical line chart: Frequency vs Sales. Points at (8, 2), (10, 22), (12, 18), (15, 4), (20, 4)]

Give **one** similarity and **one** difference in the sales of Peter and Georgina.

Similarity ....

....

Difference ....

....

(2)

(Total for Question 3 is 10 marks)

4 Kelsi rolled a dice 10 times.

Here are her scores.

1 5 6 4 4 2 2 3 4 3

(a) Find the mode.

....
(1)

(b) Work out the mean.

....
(2)

(c) Work out the range.

....
(2)

(Total for Question 4 is 5 marks)

5   Here is a list of the fruit 25 people liked best.

cherries	strawberries	cherries	rasberries	rasberries	strawberries	plums
rasberries	cherries	strawberries	plums	rasberries	rasberries	
rasberries	cherries	cherries	plums	strawberries	strawberries	
plums	rasberries	strawberries	strawberries	plums	strawberries	

(a) Complete the table for the information in the list.

Fruit	Tally	Frequency
cherries		
plums		
rasberries		
strawberries		

(2)

(b) Draw a suitable diagram to show this information in the table.
    Use the grid below.

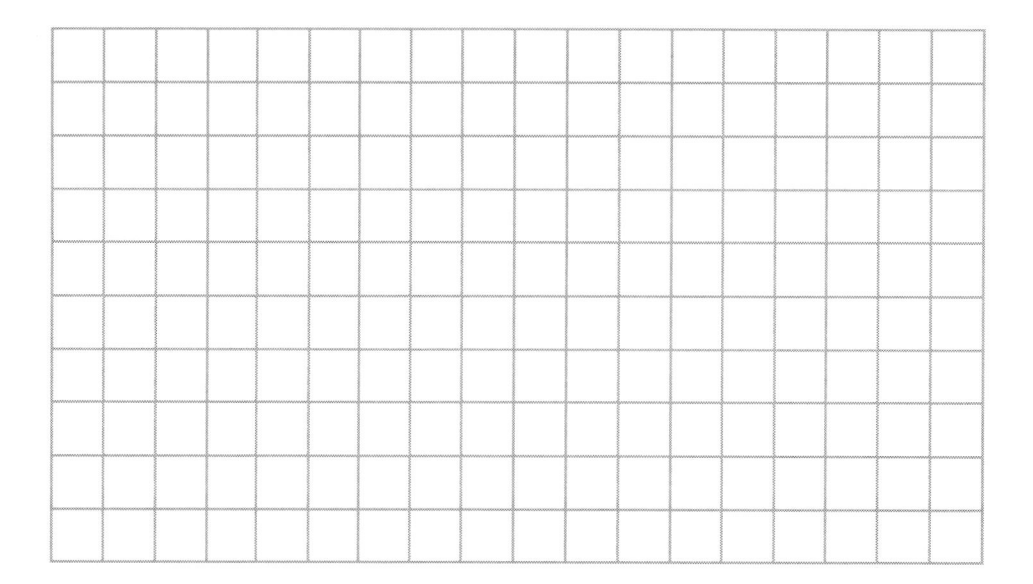

(3)

(Total for Question 5 is 5 marks)

6

	Male	Female
First year	397	608
Second year	250	210

The table gives information about the numbers of students in the two years of a college course.

Hanna wants to interview some of these students.
She takes a random sample of 50 students stratified by year and by gender.
Work out the number of students in the sample who are male and in the first year.

....

(Total for Question 6 is 3 marks)

7 Tara carried out a survey of the number of school dinners 34 students had in one week.

The table shows this information.

Number of school dinners	Frequency	
0	0	
1	8	
2	12	
3	7	
4	5	
5	2	

Calculate the mean.

....

(Total for Question 7 is 3 marks)

8 Sophie asked 32 women about the number of children they each had.

The table shows information about her results.

Number of children	Frequency	
0	9	
1	6	
2	7	
3	8	
4	2	
more than 4	0	

(a) Find the mode.

....
(1)

(b) Calculate the mean.

....
(3)

(Total for Question 8 is 4 marks)

9 The table shows some information about the ages, in years, of 60 people.

Age (in years)	Frequency
0 to 9	6
10 to 19	13
20 to 29	12
30 to 39	9
40 to 49	7
50 to 59	4
60 to 69	9

(a) Write down the modal class.

....
(1)

David says

'The median lies in the class 30 to 39'

David is wrong.

(b) Explain why.

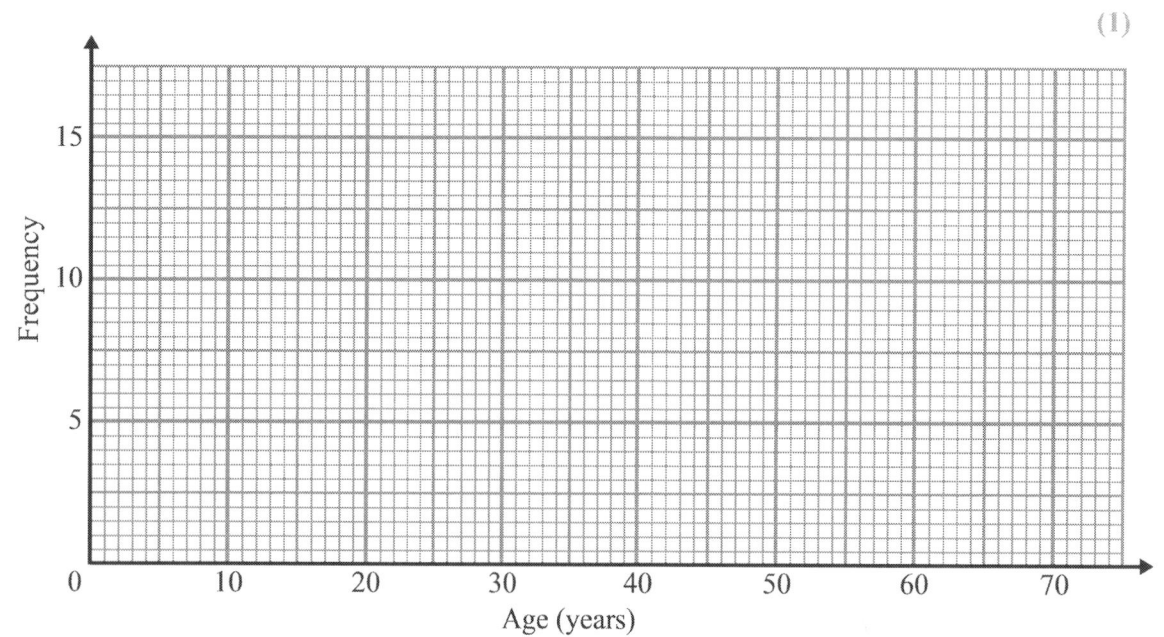

....

....
(1)

(c) On the grid, draw a frequency polygon for the information in the table.
(2)

(Total for Question 9 is 4 marks)

10 60 students take a maths test.

The test is marked out of 50.

This table shows information about the students' marks.

Maths mark	0–10	11–20	21–30	31–40	41–50
Frequency	5	13	17	19	6

On the grid, draw a frequency polygon to show this information.

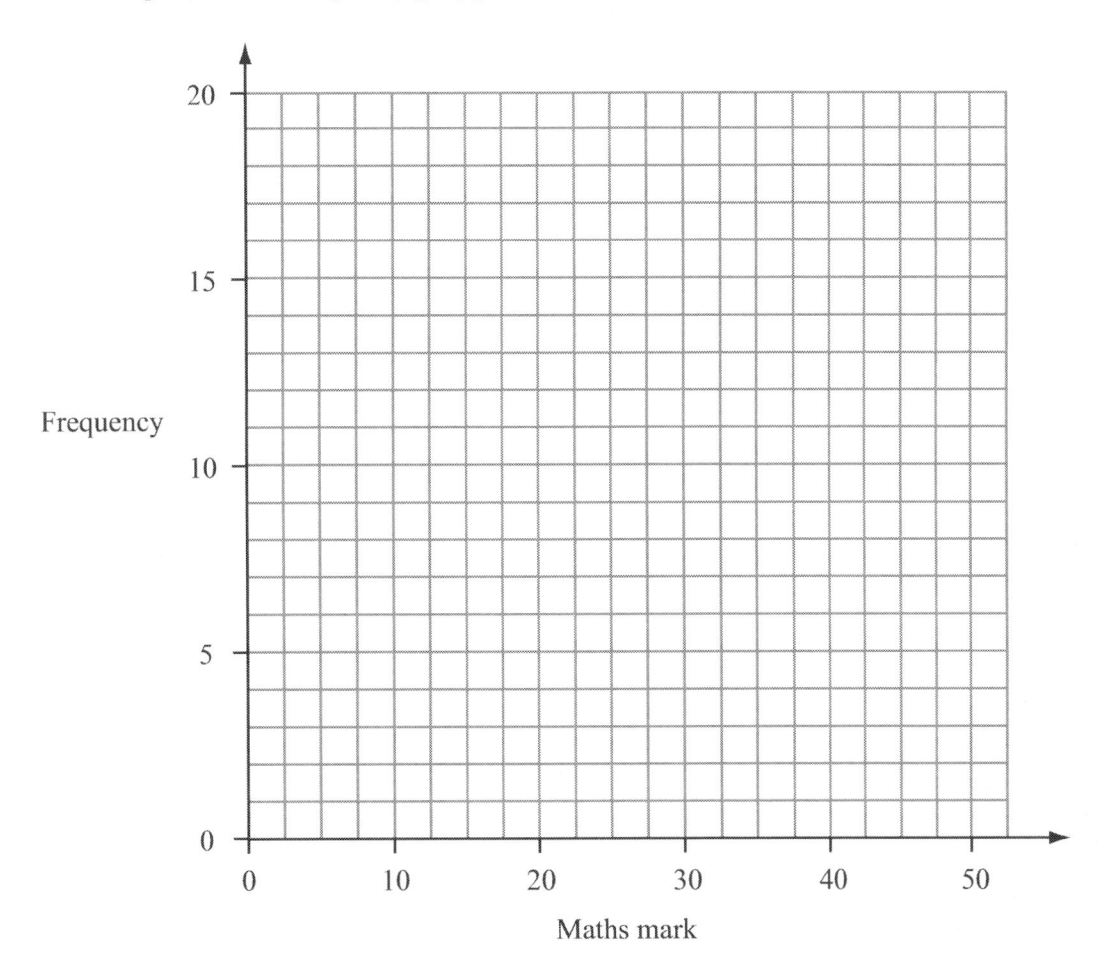

(Total for Question 10 is 2 marks)

11. The table shows some information about the weights, in kg, of 100 boxes.

Weight of box (w kg)	Frequency
$0 < w \leqslant 4$	11
$4 < w \leqslant 8$	16
$8 < w \leqslant 12$	29
$12 < w \leqslant 16$	26
$16 < w \leqslant 20$	20

Draw a frequency polygon to show this information.

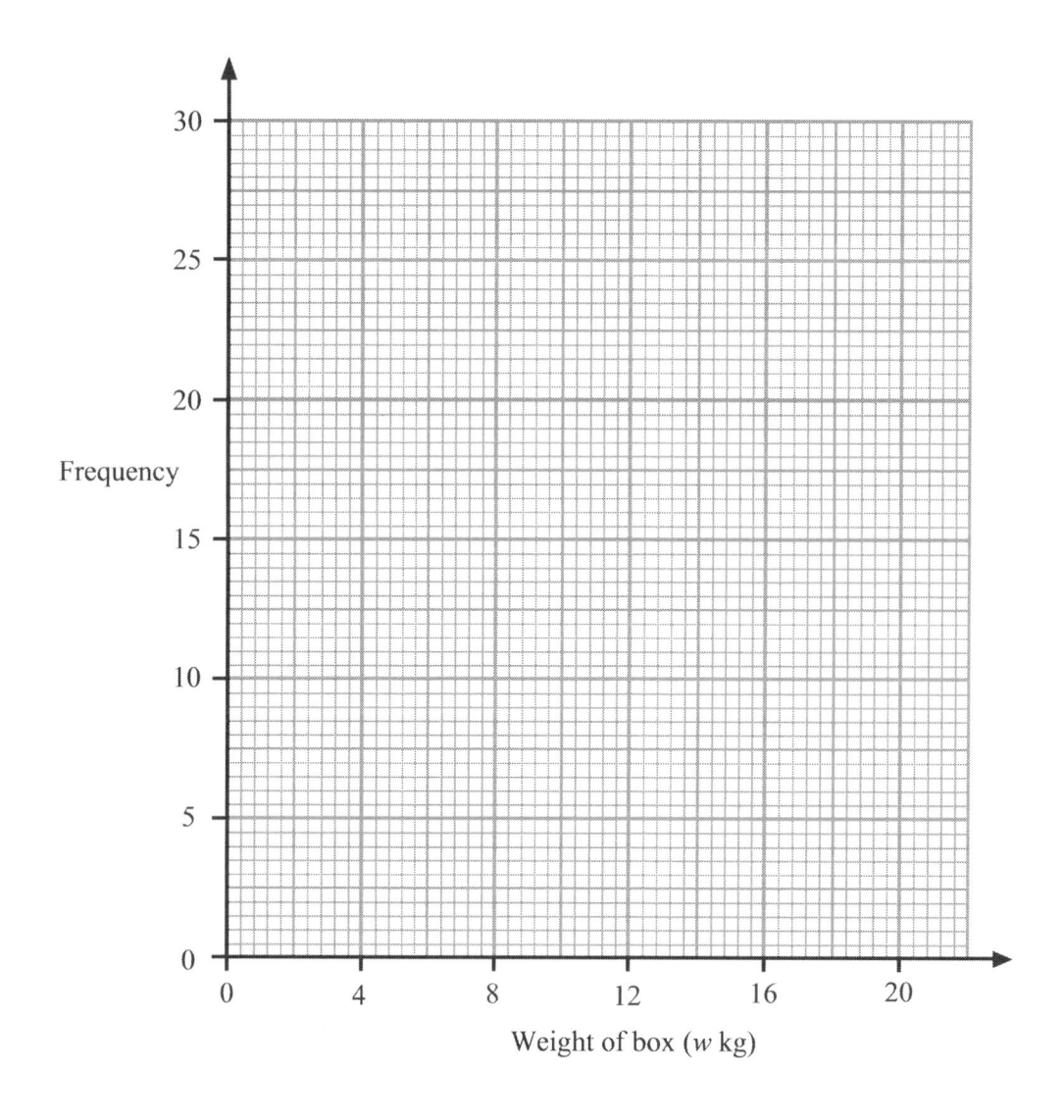

(Total for Question 11 is 2 marks)

12 The frequency table gives information about the times it took some children to get to school one day.

Time (t minutes)	Frequency
$0 < t \leq 10$	4
$10 < t \leq 20$	8
$20 < t \leq 30$	14
$30 < t \leq 40$	16
$40 < t \leq 50$	6
$50 < t \leq 60$	2

(a) Draw a frequency polygon for this information.

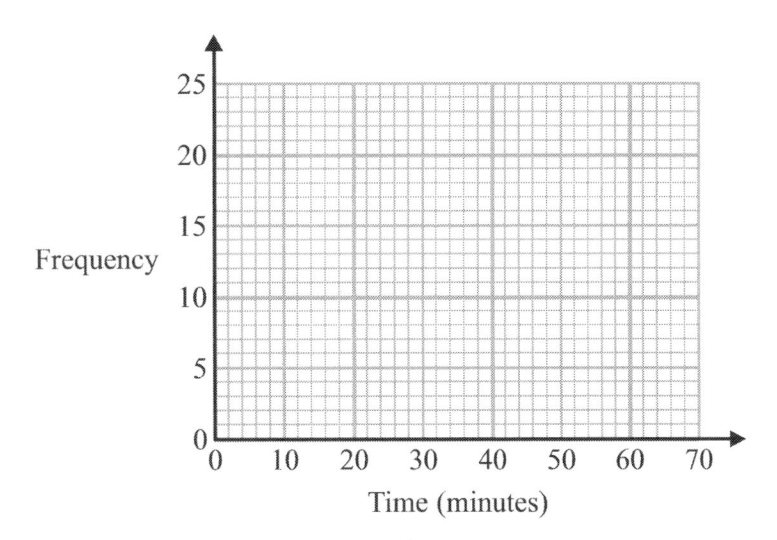

(2)

(b) Write down the modal class interval.

....
(1)

One of the children is chosen at random.

(c) Work out the probability that this child took more than 40 minutes to get to school.

....
(2)

(Total for Question 12 is 5 marks)

13 The bar chart gives information about the numbers of students in the four Year 11 classes at Cranford Academy.

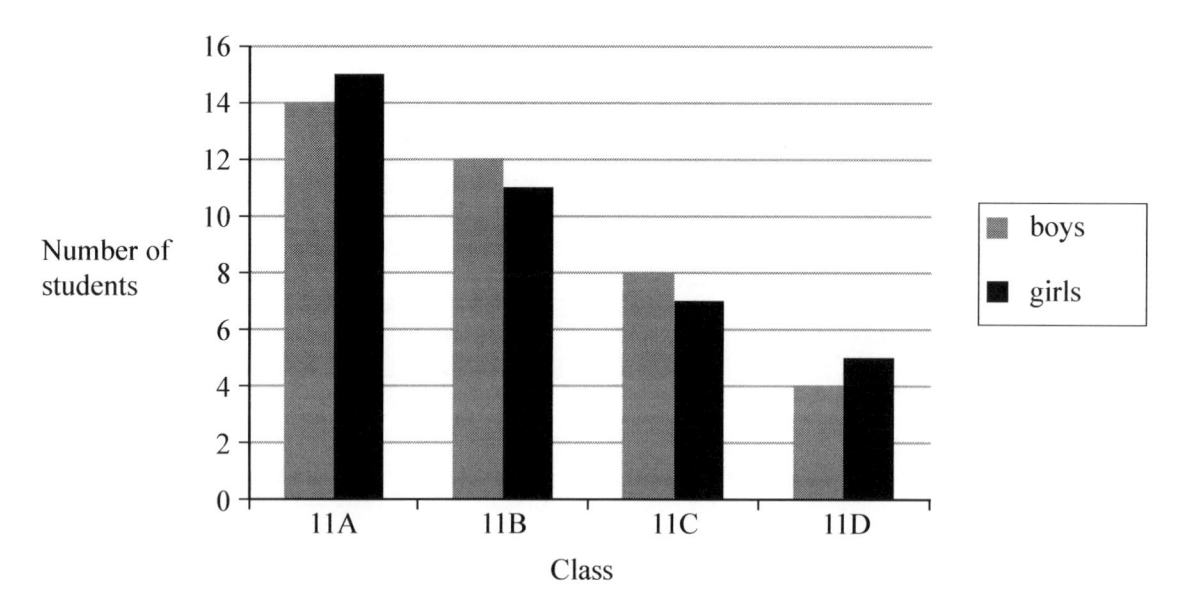

(a) What fraction of the students in class 11A are girls?

....
(2)

Sheila says,

"There are more boys than girls in Year 11 in Cranford Academy."

(b) Is Sheila correct?
You must give a reason for your answer.

(2)

The pie chart gives information about the 76 students in the same four Year 11 classes at Cranford Academy.

Number of students in Year 11 of Cranford Academy

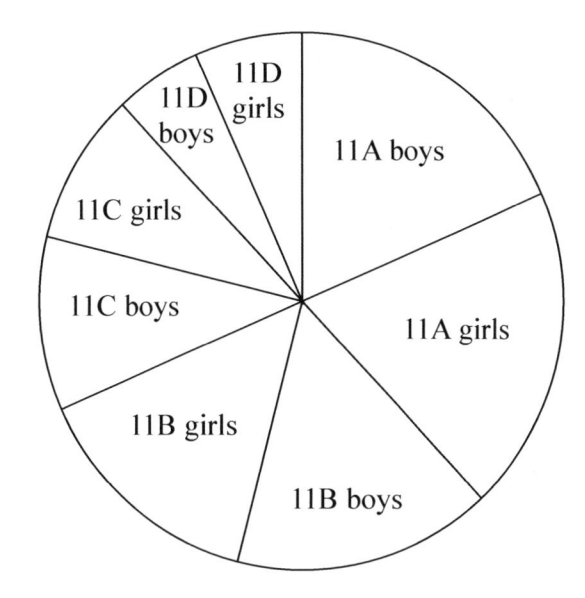

Ron says,

"It is more difficult to find out the numbers of students in each class from the pie chart than from the bar chart."

(c) Is Ron correct?
   You must give a reason for your answer.

....

....

(1)

(Total for Question 13 is 5 marks)

14 The table shows some information about the foot lengths of 40 adults.

Foot length (f cm)	Number of adults
$16 \leqslant f < 18$	3
$18 \leqslant f < 20$	6
$20 \leqslant f < 22$	10
$22 \leqslant f < 24$	12
$24 \leqslant f < 26$	9

(a) Write down the modal class interval.

....
(1)

(b) Calculate an estimate for the mean foot length.

.... cm
(3)

(Total for Question 14 is 4 marks)

15 Khalil and Haziq work in a shop from Monday to Friday.

Khalil draws a graph to show the number of TVs they each sell.

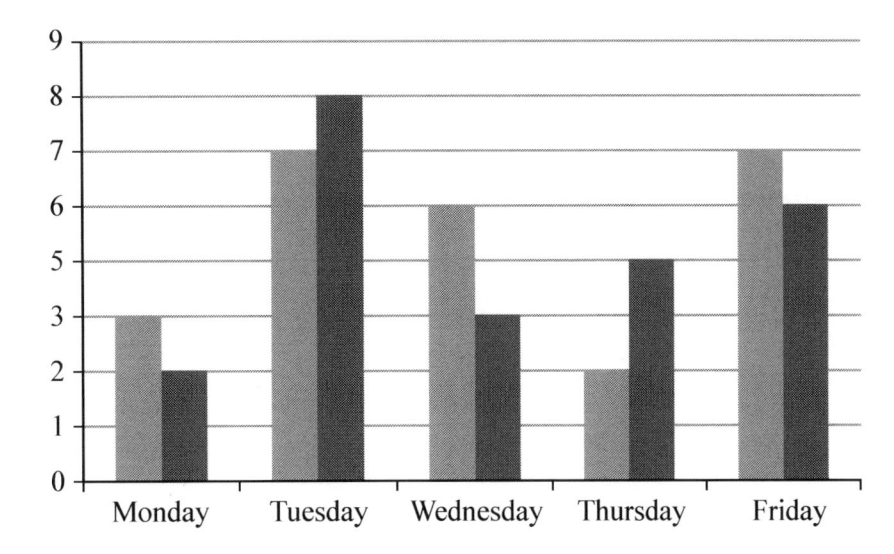

Write down **three** things that are wrong with this graph.

1 ....

....

2 ....

....

3..

....

(Total for Question 15 is 3 marks)

16 Here is a list of all the coins in Tara's purse.

£1	5p	20p	1p
20p	1p	10p	£1
20p	10p	£1	20p
10p	20p	20p	5p

Complete the table for this information.

Coin	Tally	Frequency
£1		
50p		
20p		
10p		
5p		
2p		
1p		

(Total for Question 16 is 2 marks)

17 Here are the test marks for the students in Mrs Illidge's maths class.

28	36	25	43	18	39	30
36	27	31	33	14	22	36
23	12	38	36	40	45	27

(a) Complete the frequency table for these marks.

Mark	Tally	Frequency
10 – 19		
20 – 29		
30 – 39		
40 – 49		

(3)

(b) Work out the number of these students getting less than 20 marks.

....
(1)

(Total for Question 17 is 4 marks)

18 The table shows information about the marks of 30 students in a test.

Mark	Frequency
14	2
15	10
16	2
17	3
18	13
	Total = 30

Students who scored less than the mean mark have to retake the test.

How many students have to retake the test?

You **must** show your working.

....

(**Total for Question 17 is 3 marks**)

19 The pictogram shows how some passengers spent most of their time on a flight.

Reading	▦ ▦
Watching films	▦ ▦ ▫
Listening to music	
Playing games	▦ ▦ ▯
Other	▦

Key: ▦ represents 40 people

(a) How many passengers spent most of their time playing games?

(a) ....
(1)

(b) How many **more** passengers spent most of their time watching films than reading?

(b) ....
(1)

(c) There were 360 passengers on the plane.

Complete the pictogram for listening to music.

(3)

(Total for Question 19 is 5 marks)

20 Lauren asked some people about their favourite type of holiday.
The pictogram shows her results.

Beach	□ □ □
Walking	□ □ ▯
Cruising	□ □ □ □ □ ▯
Adventure	□ □ □ □
Sightseeing	
Other	□ □ □ ▯

Key : □ represents 4 people.

(a) How many people answered Beach?

(a) ....

(1)

(b) 10 people answered Sightseeing.

Show this on the pictogram.

(1)

(c) How many **more** people answered Cruising than Other?

(c) ....

(1)

(d) How many people were asked altogether?

(d) ....

(2)

(Total for Question 20 is 5 marks)

21 Terry works in a cafe.

At noon one day he records the number of customers sitting at each table in the cafe.

Here are his results.

Number of customers sitting at a table	Number of tables
0	4
1	5
2	10
3	7
4	3
5	1

(a) Work out the total number of tables in the cafe.

(1)

(b) Work out the total number of customers sitting at tables in the cafe.

(2)

(c) Work out the mean number of customers sitting at a table.

(2)

(Total for Question 21 is 5 marks)

Write your name here

Surname

Other names

In the style of:
**Pearson Edexcel
Level 1/Level 2 GCSE (9 - 1)**

Centre Number

Candidate Number

Mathematics
Grade 9 type questions

Higher Tier

GCSE style questions arranged by topic

Paper Reference
1MA1/1H

You must have: Ruler graduated in centimetres and millimetres, protractor, pair of compasses, pen, HB pencil, eraser.

Total Marks

Instructions

- Use **black** ink or ball-point pen.
- **Fill in the boxes** at the top of this page with your name, centre number and candidate number.
- Answer **all** questions.
- Answer the questions in the spaces provided
  – there may be more space than you need.
- **Calculators may not be used.**
- Diagrams are **NOT** accurately drawn, unless otherwise indicated.
- You must **show all your working out**.

Information

- The total mark for this paper is
- The marks for **each** question are shown in brackets
  – use this as a guide as to how much time to spend on each question.

Advice

- Read each question carefully before you start to answer it.
- Keep an eye on the time.
- Try to answer every question.
- Check your answers if you have time at the end.

Turn over ▶

© Peter Bland

1 Solve the equation $\quad \dfrac{x}{2} - \dfrac{2}{x+1} = 1$

....
(Total for Question 1 is 4 marks)

2. The diagram shows a solid wax cylinder.

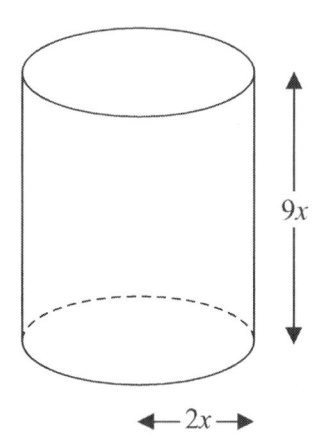

Diagram **NOT** accurately drawn

The cylinder has base radius $2x$ and height $9x$.

The cylinder is melted down and made into a sphere of radius K

Find an expression for K in terms of x.

....

(Total for Question 2 is 3 marks)

3

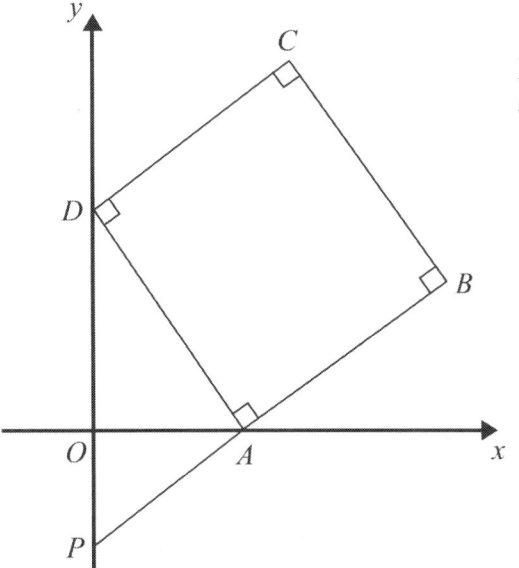

Diagram NOT accurately drawn

ABCD is a square.
P and *D* are points on the *y*-axis.
A is a point on the *x*-axis.
PAB is a straight line.

The equation of the line that passes through the points *A* and *D* is $y = -2x + 5$

Find the length of *PD*.

....

(Total for Question 3 is 4 marks)

4

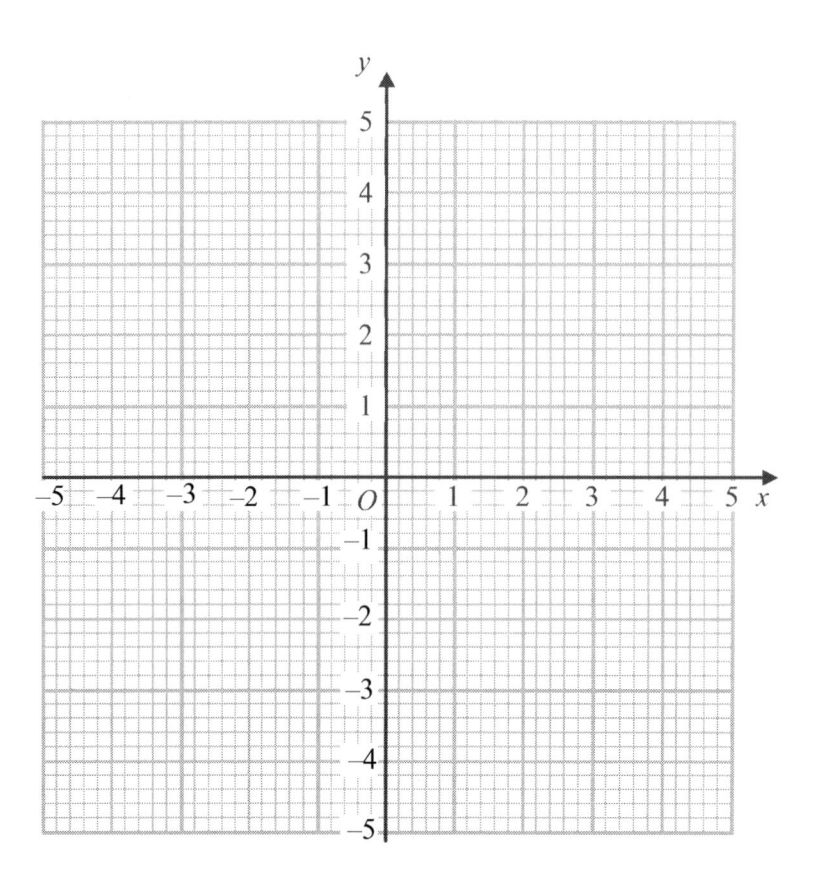

(a) On the grid, draw the graph of $x^2 + y^2 = 4$

(2)

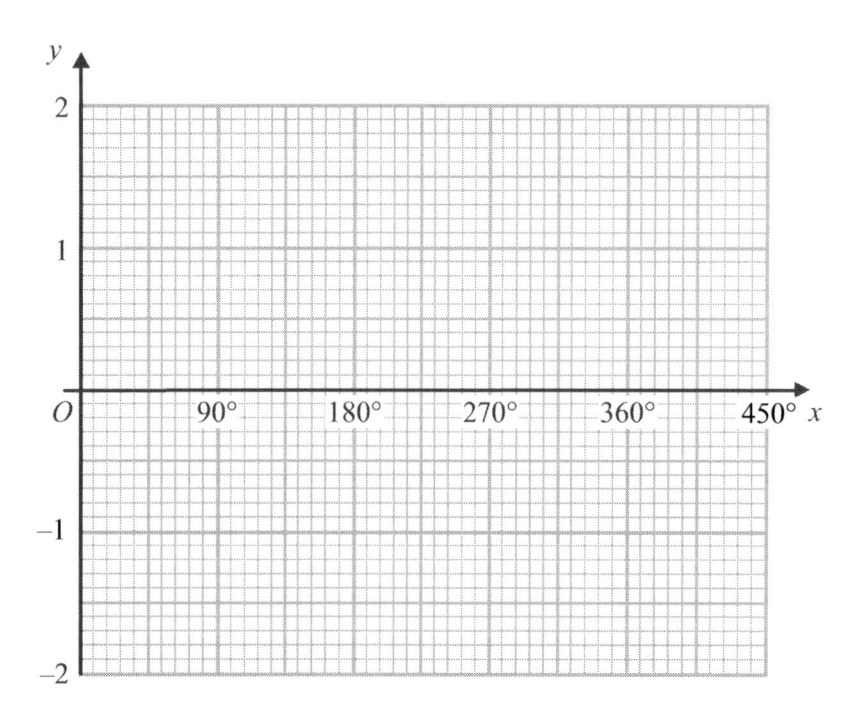

(b) On the grid, sketch the graph of $y = \cos x$ for $0° \leqslant x \leqslant 360°$

(Total for Question 4 is 4 marks)

5

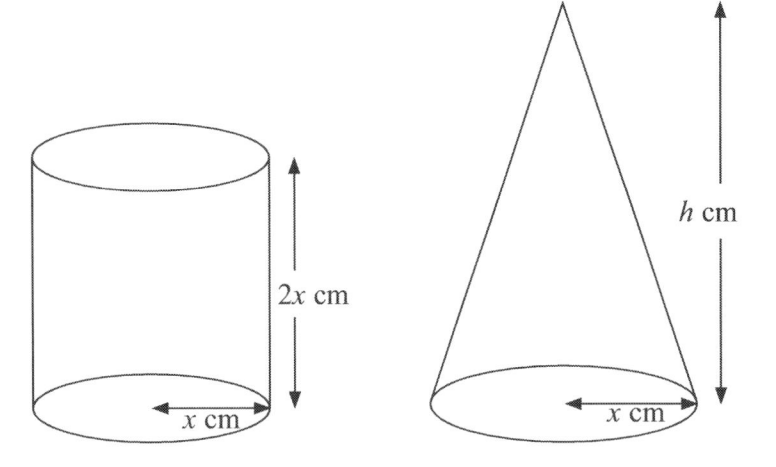

Diagrams **NOT** accurately drawn

A cylinder has base radius x cm and height $2x$ cm.

A cone has base radius x cm and height h cm.

The volume of the cylinder and the volume of the cone are equal.

Find h in terms of x.
Give your answer in its simplest form.

$h = $ ....

(Total for Question 5 is 3 marks)

6
$$\frac{1}{u} + \frac{1}{v} = \frac{1}{f}$$

$u = 2\frac{1}{2}, v = 3\frac{1}{3}$

(a) Find the value of f.

....
(3)

(b) Rearrange $\frac{1}{u} + \frac{1}{v} = \frac{1}{f}$

to make u the subject of the formula.

Give your answer in its simplest form.

....
(2)

(Total for Question 6 is 5 marks)

7

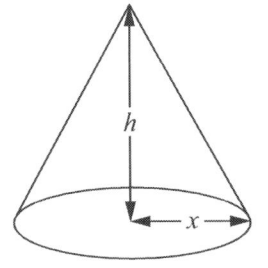

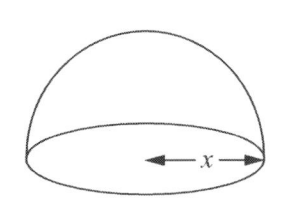

Diagram **NOT** accurately drawn

The diagram shows a solid cone and a solid hemisphere.

The cone has a base of radius x cm and a height of h cm.
The hemisphere has a base of radius x cm.
The surface area of the cone is equal to the surface area of the hemisphere.

Find an expression for h in terms of x.

....
(Total for Question 7 is 4 marks)

8

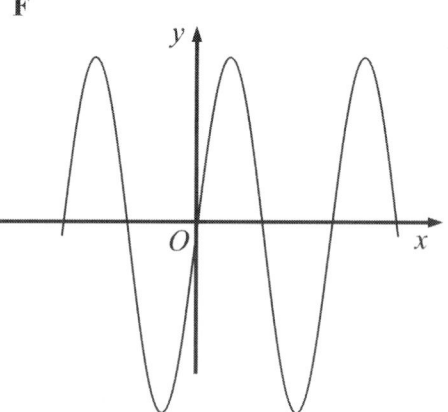

A, B, C, D, E, F (graphs)

Each equation in the table represents one of the graphs **A** to **F**.

Write the letter of each graph in the correct place in the table.

Equation	Graph
$y = 4 \sin x°$	
$y = 4 \cos x°$	
$y = x^2 - 4x + 5$	
$y = 4 \times 2^x$	
$y = x^3 + 4$	
$y = \dfrac{4}{x}$	

(Total for Question 8 is 3 marks)

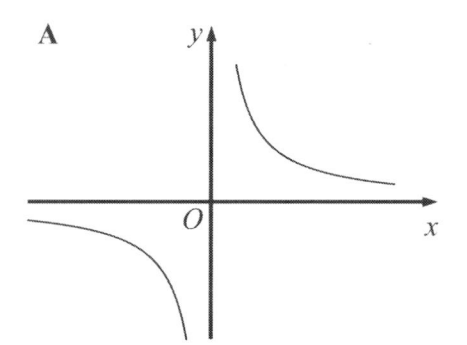

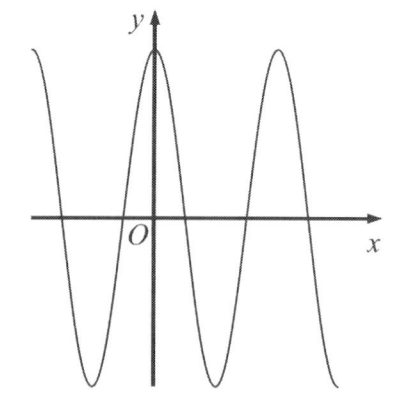

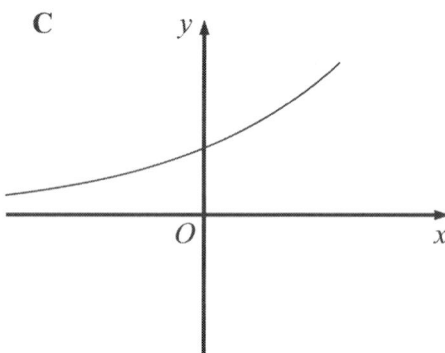

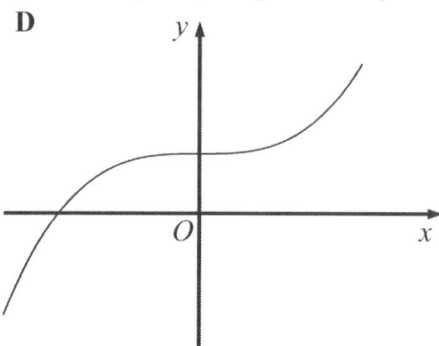

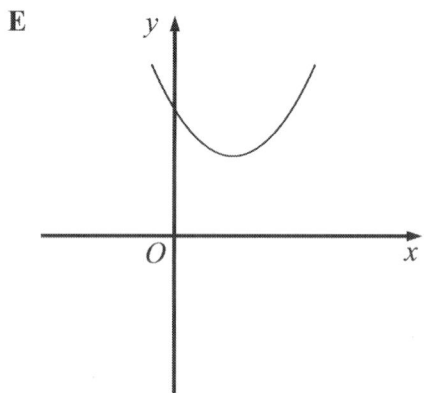

9 Here is a shape ABCDE.

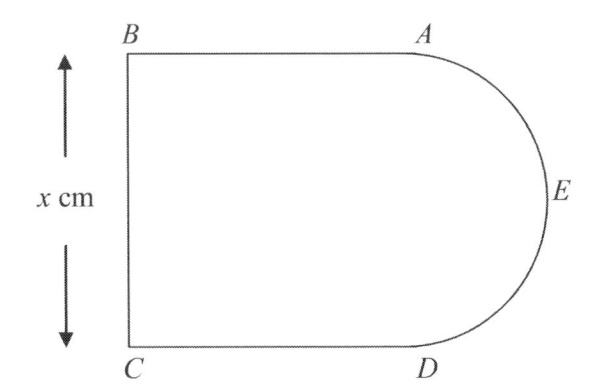

Diagram **NOT** accurately drawn

AB, BC and CD are three sides of a square.
BC = x cm.
AED is a semicircle with diameter AD.

The perimeter, P cm, of the shape ABCDE is given by the formula

$$P = 3x + \frac{\pi x}{2}$$

(a) Rearrange this formula to make x the subject.

....

(2)

The area, A cm^2, of this shape is given by $A = kx^2$ where k is a constant.

(b) Find the exact value of k.
    Give your answer in its simplest form.

....
(3)

(Total for Question 9 is 5 marks)

10   Express the recurring decimal $0.2\dot{1}\dot{3}$ as a fraction.

....

(Total for Question 10 is 3 marks)

11

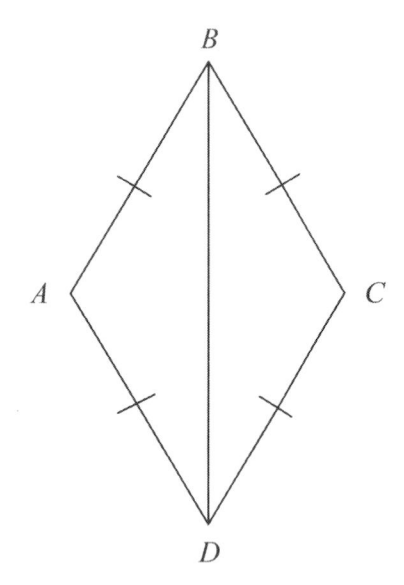

Diagram **NOT** accurately drawn

In the diagram, $AB = BC = CD = DA$.

Prove that triangle ADB is congruent to triangle CDB.

(Total for Question 11 is 3 marks)

12 Prove, using algebra, that the sum of two consecutive whole numbers is always an odd number.

(Total for Question 12 is 3 marks)

13 The table shows information about the ages, in years, of 1000 teenagers.

Age (years)	13	14	15	16	17	18	19
Number of teenagers	158	180	165	141	131	115	110

Sophie takes a sample of 50 of these teenagers, stratified by age.

Calculate the number of 14 year olds she should have in her sample.

....

(Total for Question 13 is 2 marks)

---

14 P is inversely proportional to V.

When $V = 8$, $P = 5$

(a) Find a formula for P in terms of V.

$P = $ ....

(3)

(b) Calculate the value of P when $V = 2$

....

(1)

(Total for Question 14 is 4 marks)

15

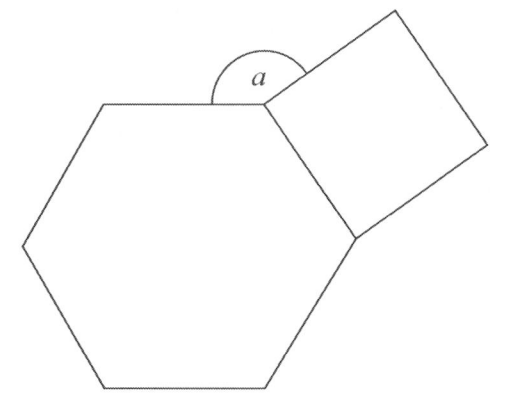

Diagram **NOT** accurately drawn

The diagram shows a regular hexagon and a square.

Calculate the size of the angle a.

.... °

(Total for Question 15 is 4 marks)

16

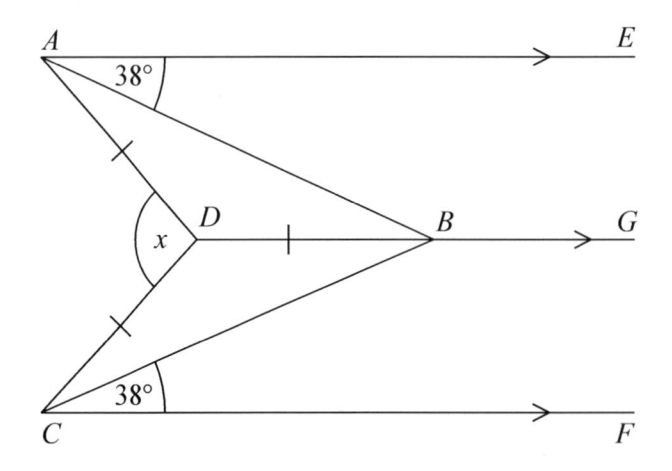

AE, *DBG* and *CF* are parallel.
DA = *DB* = *DC*.
Angle *EAB* = angle *BCF* = 38°

Work out the size of the angle marked *x*.
You must show your working.

.... °

(Total for Question 16 is 3 marks)

17 $A(-2, 1)$, $B(6, 5)$ and $C(4, k)$ are the vertices of a right-angled triangle ABC.
Angle ABC is the right angle.

Find an equation of the line that passes through A and C.
Give your answer in the form $ay + bx = c$ where a, b and c are integers.

....

(Total for Question 17 is 5 marks)

18 Here is a speed-time graph for a car journey.
The journey took 100 seconds.

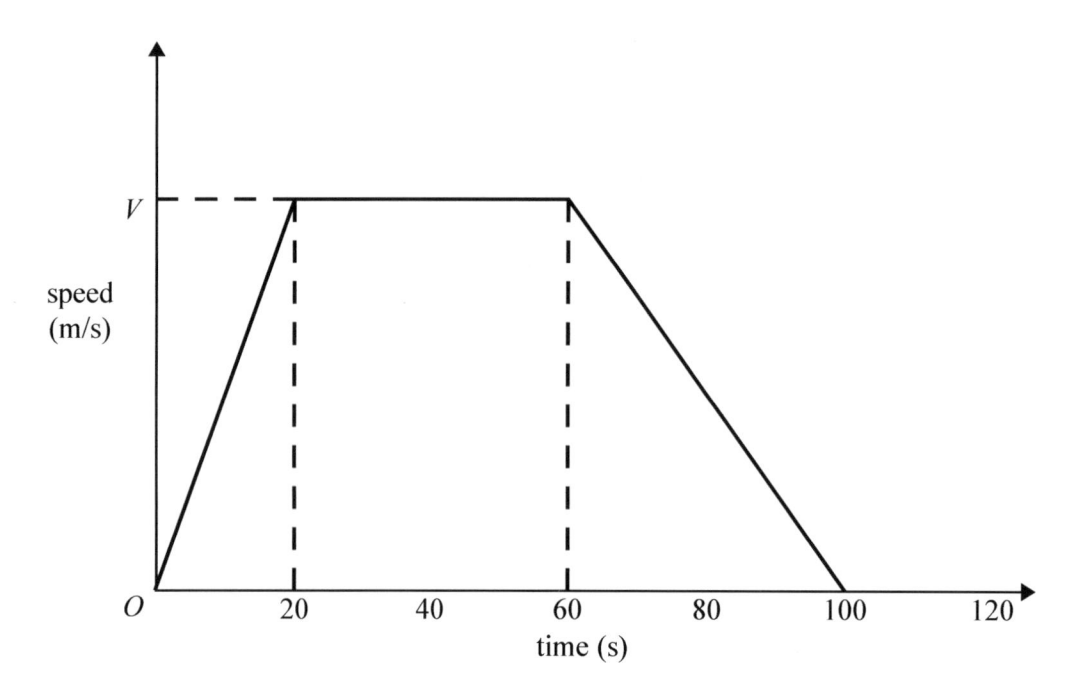

The car travelled 1.75 km in the 100 seconds.

(a) Work out the value of V.

....
(3)

(b) Describe the acceleration of the car for each part of this journey.

....

....

....

....
(2)

(Total for Question 18 is 5 marks)

19 In this question all dimensions are in centimetres.

A solid has uniform cross section.

The cross section is a rectangle and a semicircle joined together.

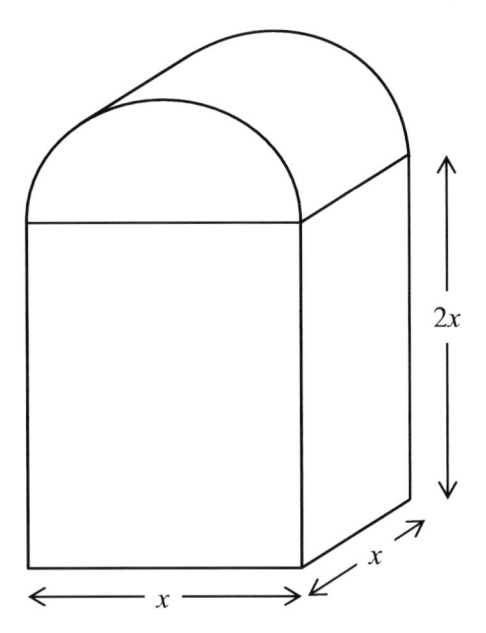

Work out an expression, in cm³, for the **total** volume of the solid.

Write your expression in the form $ax^3 + \frac{1}{b}\pi x^3$ where a and b are integers.

.... cm³

(Total for Question 19 is 4 marks)

20 f(x) = 2x + c

g(x) = cx + 5

fg(x) = 6x + d

c and d are constants.

Work out the value of d.

....

(Total for Question 20 is 3 marks)

Write your name here

Surname

Other names

In the style of:
**Pearson Edexcel
Level 1/Level 2 GCSE (9 - 1)**

Centre Number

Candidate Number

Mathematics
Grade 9 type questions

Higher Tier

GCSE style questions arranged by topic

Paper Reference
1MA1/2H

You must have: Ruler graduated in centimetres and millimetres, protractor, pair of compasses, pen, HB pencil, eraser, calculator.

Total Marks

Instructions

- Use **black** ink or ball-point pen.
- **Fill in the boxes** at the top of this page with your name, centre number and candidate number.
- Answer **all** questions.
- Answer the questions in the spaces provided
  – there may be more space than you need.
- **Calculators may be used.**
- If your calculator does not have a π button, take the value of π to be 3.142 unless the question instructs otherwise.
- Diagrams are **NOT** accurately drawn, unless otherwise indicated.
- You must **show all your working out**.

Information

- The total mark for this paper is
- The marks for **each** question are shown in brackets
  – use this as a guide as to how much time to spend on each question.

Advice

- Read each question carefully before you start to answer it.
- Keep an eye on the time.
- Try to answer every question.
- Check your answers if you have time at the end.

Turn over

© Peter Bland

1   258 Year 9 were choosing the subjects they would be taking in Year 10. The table shows information about these students.

	Subject to be studied		
	Geography	History	Spanish
Male	45	52	26
Female	25	48	62

A sample, stratified by the subject studied and by gender, of 50 of the 258 students is taken.

(a) Work out the number of male students studying Spanish in the sample.

....
(2)

(b) Work out the number of female students in the sample.

....
(2)

(Total for Question 1 is 4 marks)

2   Prove that $(3x + 1)^2 - (3x - 1)^2$ is a multiple of 4, for all positive integer values of x.

(Total for Question 2 is 3 marks)

3

The diagram shows an equilateral triangle ABC with sides of length 6 cm.

P is the midpoint of AB.
Q is the midpoint of AC.
APQ is a sector of a circle, centre A.

Calculate the area of the shaded region.
Give your answer correct to 3 significant figures.

.... cm²

(Total for Question 3 is 4 marks)

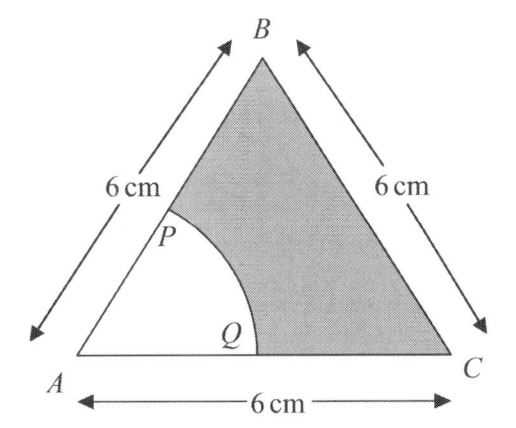

4   Make *A* the subject of the formula

$$x = \sqrt{\frac{A}{3}}$$

A = ...

(Total for Question 4 is 2 marks)

5   (a) Write 12 500 in standard form.

....
(1)

(b) Write 2.48×10^{-3} as an ordinary number.

....
(1)

(c) Work out the value of

$$23\,500 \div (1.25 \times 10^{-4})$$

Give your answer in standard form.

....
(2)

(Total for Question 5 is 4 marks)

6   **X** and **Y** are two solid shapes which are mathematically similar.
    The shapes are made from the same material.

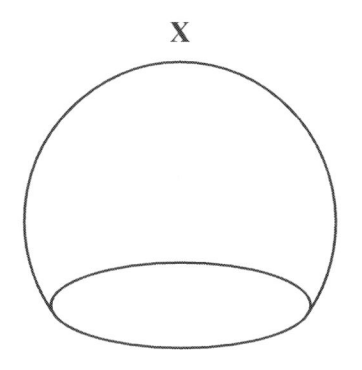

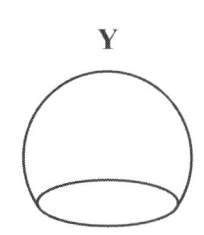

                                                    Diagram **NOT**
                                                    accurately drawn

    The surface area of **X** is 50 cm².
    The surface area of **Y** is 18 cm².
    The mass of **X** is 500 grams.

    Calculate the mass of **Y**.

    .... grams

    (Total for Question 6 is 4 marks)

7 The diagram shows a sector of a circle with centre O.
   The radius of the circle is 8 cm.

   XYZ is an arc of the circle.
   XZ is a chord of the circle.
   Angle $XOZ = 40°$

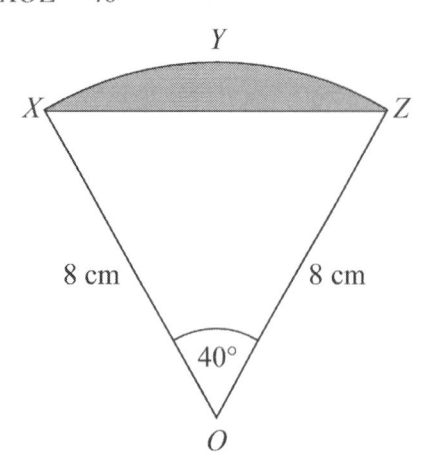

   Diagram **NOT** accurately drawn

   Calculate the area of the shaded segment.
   Give your answer correct to 3 significant figures.

   .... cm²

   (Total for Question 7 is 5 marks)

8 The table shows six expressions.
x is a positive integer.

$2x - 3$	$3x - 2$	$3(x + 4)$	$4x + 1$	$4(3x + 1)$	$2x + 1$

(a) From the table, write the expression whose value is

   (i) always even

   ....

   (ii) always a multiple of 3

   ....
   (2)

(b) From the table, write the expression which is a factor of $4x^2 - 1$

   ....
   (1)

(Total for Question 8 is 3 marks)

9 (a) $n > -3$

Show this inequality on the number line.

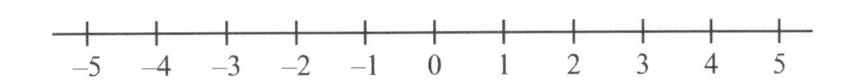

$$\begin{array}{c} \vdash\!\!\!\!\!\!\!\dashv\!\!\!\!\dashv\!\!\!\!\dashv\!\!\!\!\dashv\!\!\!\!\dashv\!\!\!\!\dashv\!\!\!\!\dashv\!\!\!\!\dashv\!\!\!\!\dashv\!\!\!\!\dashv\!\!\!\!\vdash \\ -5\ \ -4\ \ -3\ \ -2\ \ -1\ \ \ 0\ \ \ 1\ \ \ 2\ \ \ 3\ \ \ 4\ \ \ 5 \end{array}$$

(2)

(b) Solve the inequality $7x + 36 \leq 8$

....
(2)

(Total for Question 9 is 4 marks)

10 In a sale the normal price of a pen is reduced by 10%.

The sale price of the pen is £4.86

Calculate the normal price of the pen.

£..

(Total for Question 10 is 3 marks)

11 The diagram shows two similar triangles.

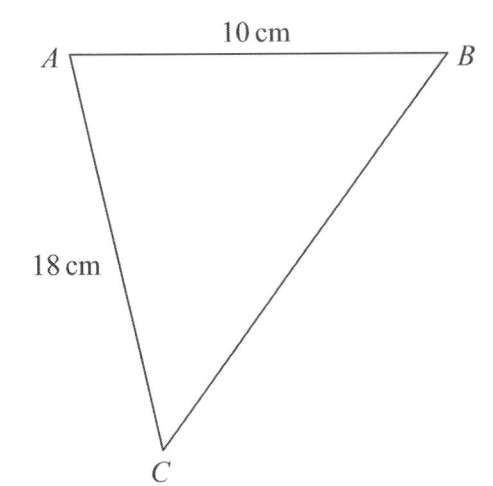

Diagram **NOT** accurately drawn

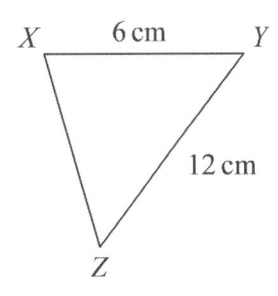

In triangle ABC, AB = 10 cm and AC = 18 cm.
In triangle XYZ, XY = 6 cm and YZ = 12 cm.

Angle ABC = angle XYZ.
Angle CAB = angle ZXY.

(a) Calculate the length of BC.

.... cm
(2)

(b) Calculate the length of XZ.

.... cm
(2)

(Total for Question 11 is 4 marks)

12 The surface area of Venus is 510 072 000 km².

The surface area of Jupiter is 6.21795×10^{10} km².

The surface area of Jupiter is greater than the surface area of Venus.
How many times greater?
Give your answer in standard form.

(Total for Question 12 is 5 marks)

13 The table shows some expressions.
w, x, y and z represent lengths.
π and 2 are numbers that have no dimensions.

$y^2(x+z)$	$\pi w^2 y^2$	$\dfrac{w^3 x}{y^3}$	$\pi w^2 x$	$\dfrac{2w^3 z}{y}$	z^2	$2w + x^2$

Tick (✓) the boxes underneath the **three** expressions which could represent volumes.

(Total for Question 13 is 3 marks)

14 There are three big employment sites in Knutsford.
The table shows the number of employees in each of these sites.

Barclays	**Longridge**	**Parkgate**
750	700	900

Georgina takes a sample of 50 employees stratified by site. Work out the number of employees from Longridge in the sample.

....

(Total for Question 14 is 2 marks)

15 (a) On the number line below, show the inequality $-2 < x < 3$

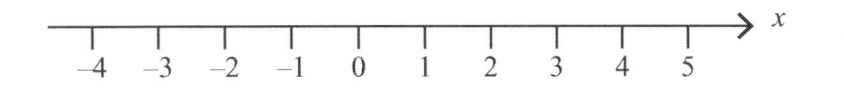

(1)

(b) Here is an inequality, in y, shown on a number line.

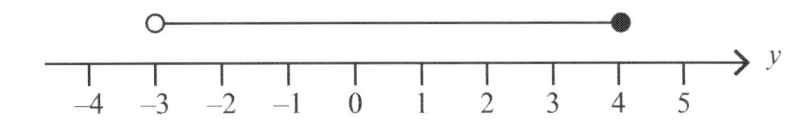

Write down the inequality.

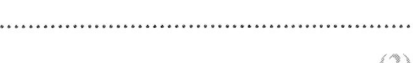

....

(2)

(c) Solve the inequality $4t - 5 > 9$

....

(2)

(Total for Question 15 is 5 marks)

16

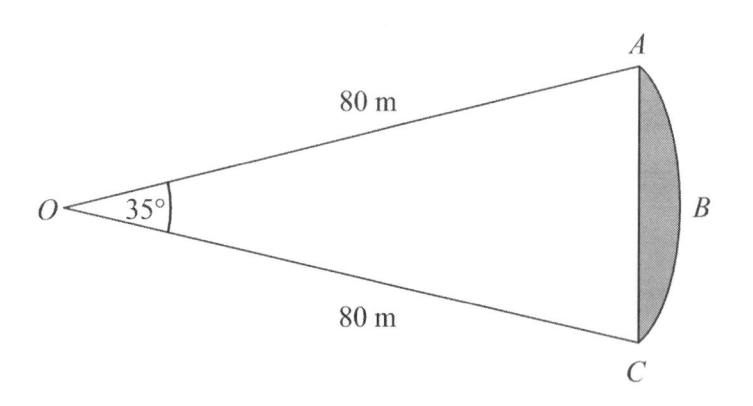

Diagram **NOT** accurately drawn

ABC is an arc of a circle centre *O* with radius 80 m.
AC is a chord of the circle.
Angle *AOC* = 35°.

Calculate the area of the shaded region.
Give your answer correct to 3 significant figures.

.... m²

(Total for Question 16 is 5 marks)

17 The table below gives some information about some students in a school.

Year group	Boys	Girls	Total
Year 12	126	94	220
Year 13	77	85	162
Total	203	179	382

Andrew is going to carry out a survey of these students.
He uses a sample of 50 students, stratified by year group and gender.

Work out the number of Year 13 girls that should be in his sample.

....

(Total for Question 17 is 2 marks)

---

18 y is directly proportional to x.

When $x = 500$, $y = 10$

(a) Find a formula for y in terms of x.

$y = $ ....
(3)

(b) Calculate the value of y when $x = 350$

$y = $ ....
(1)

(Total for Question 18 is 4 marks)

19   A and B are vertices of a cuboid.

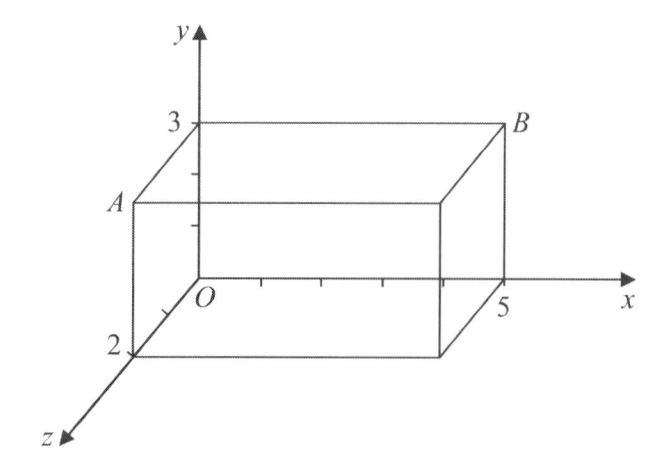

Diagram NOT accurately drawn

(a) Write down the coordinates of point A.

(.......... , ... , ...)
(1)

(b) Write down the coordinates of point B.

(.......... , ... , ...)
(1)

(Total for Question 19 is 2 marks)

20 (a) Write 83 500 000 in standard form.

....
(1)

(b) Work out $(5.2 \times 10^{-7}) \times (2.8 \times 10^{-9})$

Give your answer in standard form.

....
(2)

(Total for Question 20 is 3 marks)

21 Sheila invests £2000 in a savings account for 3 years.

The account pays compound interest at an annual rate of

  2.5% for the first year

  x% for the second year

  x% for the third year

There is a total amount of £2124.46 in the savings account at the end of 3 years.

(a) Work out the rate of interest in the second year.

(4)

Katy goes to work by train.

The cost of her weekly train ticket increases by 12.5% to £225

(b) Work out the cost of her weekly train ticket before this increase.

(2)

(Total for Question 21 is 6 marks)

22 d is inversely proportional to c

When $c = 280$, $d = 25$

Find the value of d when $c = 350$

$d = $ ....

(Total for Question 22 is 3 marks)

---

23 Prove algebraically that

$(2n + 1)^2 - (2n + 1)$ is an even number

for all positive integer values of n.

(Total for Question 23 is 3 marks)

24 In triangle *RPQ*,

   RP = 8.7 cm
   PQ = 5.2 cm
   Angle *PRQ* = 32°

(a) Assuming that angle *PQR* is an acute angle, calculate the area of triangle *RPQ*.
Give your answer correct to 3 significant figures.

..............................cm²
(4)

(b) If you did not know that angle *PQR* is an acute angle, what effect would this have on your calculation of the area of triangle *RPQ*?

....

....

....
(1)

(Total for Question 24 is 5 marks)

25 $\xi = \{1, 2, 3, 4, 5, 6, 7, 8, 9, 10, 11, 12\}$

S = square numbers

E = even numbers

(a) Complete the Venn diagram.

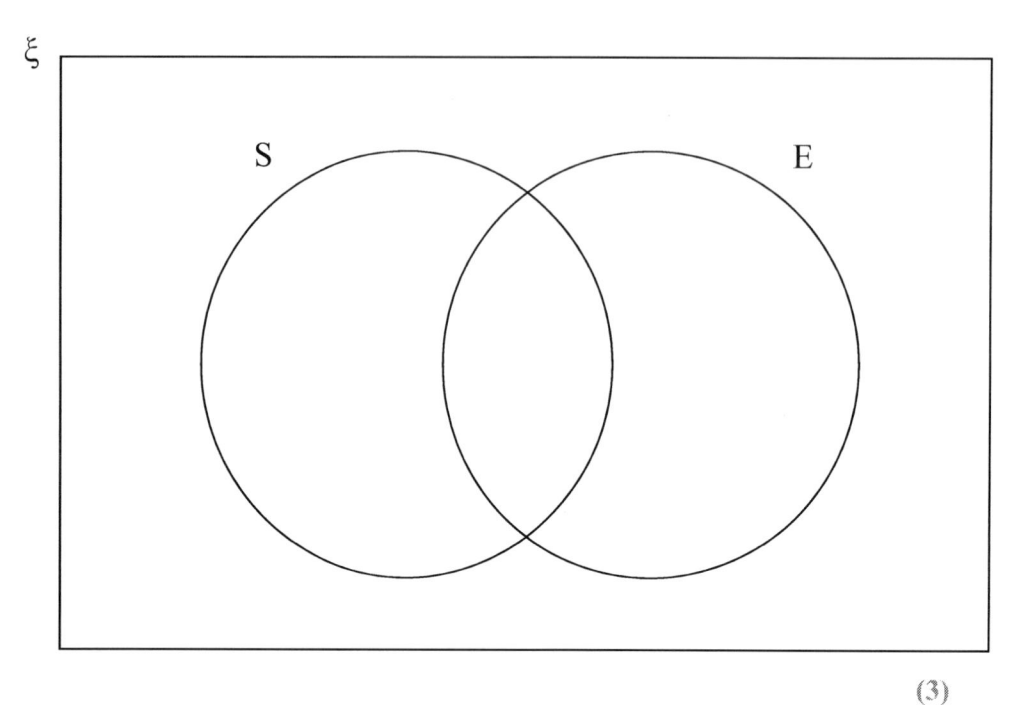

(3)

(b) One of the numbers is chosen at random.

Write down $P(S \cap E)$

....
(1)

(Total for Question 25 is 4 marks)

26 Written as the product of its prime factors

$$672 = 2^5 \times 3 \times 7$$

(a) Write 252 as the product of its prime factors.

....
(2)

(b) Work out the value of the highest common factor of 672 and 252

....
(1)

(Total for Question 26 is 3 marks)

27 (a) Write $x^2 + 10x + 29$ in the form $(x+a)^2 + b$.

(a) ....

(3)

(b) Write down the coordinates of the turning point of the graph of $y = x^2 + 10x + 29$.

(b) (................ , ...)

(1)

(Total for Question 27 is 4 marks)

Write your name here

Surname

Other names

In the style of:
Pearson Edexcel
Level 1/Level 2 GCSE (9 - 1)

Centre Number

Candidate Number

Mathematics

Grade 9 type questions

Higher Tier

GCSE style questions arranged by topic

Paper Reference
1MA1/3H

You must have: Ruler graduated in centimetres and millimetres, protractor, pair of compasses, pen, HB pencil, eraser, calculator.

Total Marks

Instructions

- Use **black** ink or ball-point pen.
- **Fill in the boxes** at the top of this page with your name, centre number and candidate number.
- Answer **all** questions.
- Answer the questions in the spaces provided
  – there may be more space than you need.
- **Calculators may be used.**
- If your calculator does not have a π button, take the value of π to be 3.142 unless the question instructs otherwise.
- Diagrams are **NOT** accurately drawn, unless otherwise indicated.
- You must **show all your working out**.

Information

- The total mark for this paper is
- The marks for **each** question are shown in brackets
  – use this as a guide as to how much time to spend on each question.

Advice

- Read each question carefully before you start to answer it.
- Keep an eye on the time.
- Try to answer every question.
- Check your answers if you have time at the end.

Turn over ▶

© Peter Bland

1 Lara asked 50 people which drinks they liked from tea, coffee and milk.

All 50 people like at least one of the drinks
19 people like all three drinks.
16 people like tea and coffee but do **not** like milk.
21 people like coffee and milk.
24 people like tea and milk.
40 people like coffee.
1 person likes only milk.

Lara selects at random one of the 50 people.

(a) Work out the probability that this person likes tea.

....

(4)

(b) Given that the person selected at random from the 50 people likes tea, find the probability that this person also likes exactly one other drink.

....

(2)

(Total for Question 1 is 6 marks)

2   Here are two function machines, **A** and **B**.

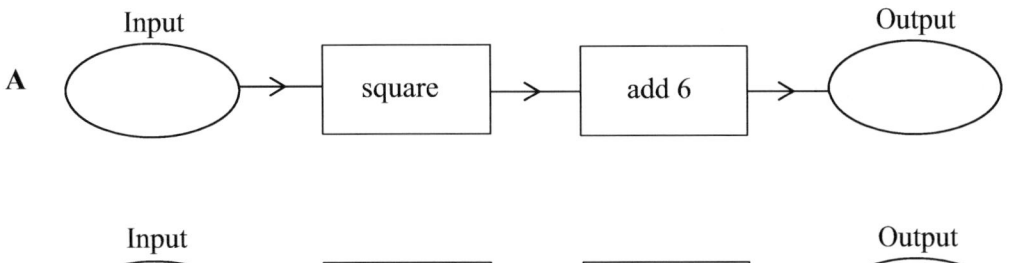

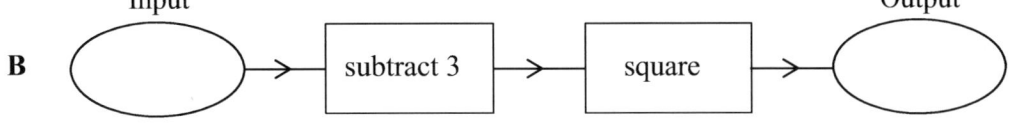

A   Input → square → add 6 → Output

B   Input → subtract 3 → square → Output

Both machines have the same input.

Work out the range of input values for which the output of **A** is **less** than the output of **B**.

....

(Total for Question 2 is 4 marks)

3 (a) A function is represented by the following function machine.

Input ⟶ × 2 ⟶ + 5 ⟶ Output

(i) A number is input into the machine.
The output is used as a new input.
The second output is 11.

Work out the number that was the **first input.**

(a)(i) ....
(2)

(ii) A number is input into the machine.
The output given is the same number.

Work out the number.

(ii) ....
(3)

(b) Another function machine is shown below.

Input ⟶ × ...... ⟶ − ...... ⟶ Output

If the Input is 3, the Output is 5.

If the Input is 7, the Output is 25.

Use this information to fill in the two boxes.

(3)

(Total for Question 3 is 8 marks)

4 Louis and Tara are investigating the growth in the population of a type of bacteria.
They have two flasks A and B.

At the start of day 1, there are 1000 bacteria in flask A.
The population of bacteria grows exponentially at the rate of 50% per day.

(a) Show that the population of bacteria in flask A at the start of each day forms a geometric progression.

(2)

The population of bacteria in flask A at the start of the 10th day is k times the population of bacteria in flask A at the start of the 6th day.

(b) Find the value of k.

....
(2)

At the start of day 1 there are 1000 bacteria in flask B.
The population of bacteria in flask B grows exponentially at the rate of 30% per day.

(c) Sketch a graph to compare the size of the population of bacteria in flask A and in flask B.

(1)

(Total for Question 4 is 5 marks)

5   An approximate solution to an equation is found using this iterative process.

$$x_{n+1} = \frac{(x_n)^3 - 3}{8} \text{ and } x_1 = -1$$

(a) Work out the values of x_2 and x_3

$x_2 = $ ....

$x_3 = $ ....

(2)

(b) Work out the solution to 6 decimal places.

$x = $ ....

(1)

(Total for Question 5 is 3 marks)

6   (a) By completing the square, find the roots of the equation

$$x^2 - 4x - 3 = 0$$

Give your answer in surd form.

(3)

(b) Show algebraically that $x^2 - 7x + 13$ has no real roots.

(3)

(Total for Question 6 is 6 marks)

7   Use an iterative formula to find the positive root of the equation

$$y = x^2 + x - 5$$

Give your answer to 5 decimal places.

x = ...
(5)

(Total for Question 7 is 5 marks)

8 Here are the first six terms of a Fibonacci sequence.

$$1 \quad 1 \quad 2 \quad 3 \quad 5 \quad 8$$

The rule to continue a Fibonacci sequence is,

the next term in the sequence is the sum of the two previous terms.

(a) Find the 9th term of this sequence.

....
(1)

The first three terms of a different Fibonacci sequence are

$$a \quad b \quad a + b$$

(b) Show that the 6th term of this sequence is $3a + 5b$

(2)

Given that the 3rd term is 7 and the 6th term is 29,

(c) find the value of a and the value of b.

....
(3)

(Total for Question 8 is 6 marks)

9 (a) Factorise $y^2 + 7y + 6$

....
(2)

(b) Solve $6x + 4 > x + 17$

....
(2)

(c) n is an integer with $-5 < 2n \leqslant 6$

Write down all the values of n

....
(2)

(Total for Question 9 is 6 marks)

10 The function f is such that

$$f(x) = 4x - 1$$

(a) Find $f^{-1}(x)$

$f^{-1}(x) = $ ....
(2)

(Total for Question 10 is 2 marks)

11 (a) Show that the equation $x^3 + 4x = 1$ has a solution between $x = 0$ and $x = 1$

(2)

(b) Show that the equation $x^3 + 4x = 1$ can be arranged to give $x = \dfrac{1}{4} - \dfrac{x^3}{4}$

(1)

(c) Starting with $x_0 = 0$, use the iteration formula $x_{n+1} = \dfrac{1}{4} - \dfrac{x_n^3}{4}$ twice, to find an estimate for the solution of $x^3 + 4x = 1$

....
(3)

(Total for Question 11 is 6 marks)

12 *ABC* is a triangle with *AB* = *AC*

BA is parallel to *CD*.

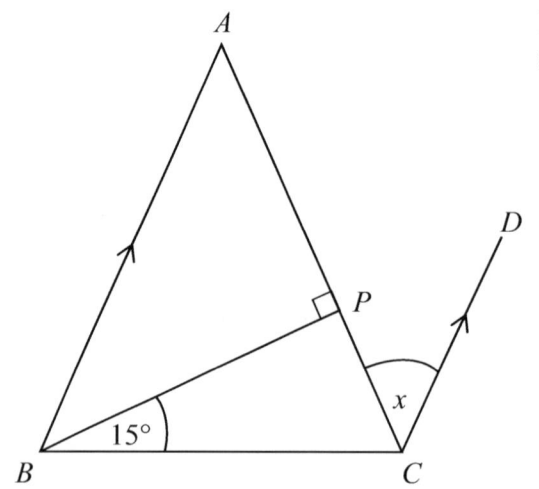

Diagram **NOT** accurately drawn

Show that angle *x* = 30°

(3)

(Total for Question 12 is 3 marks)

13 *ABCD* is a rhombus.

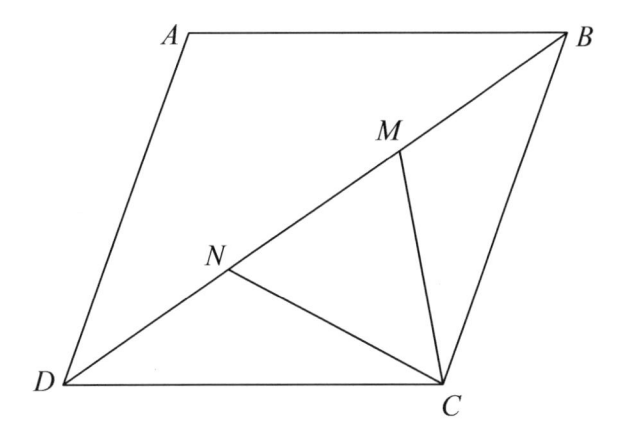

M and *N* are points on *BD* such that *DN* = *MB*.

Prove that triangle *DNC* is congruent to triangle *BMC*.

(Total for Question 13 is 3 marks)

14 The pressure at sea level is 101 325 Pascals.

Any rise of 1 km above sea level decreases the pressure by 14%

For example,
        at 3 km above sea level the pressure is 14% less than at 2 km

Work out the pressure at 4 km above sea level.

Give your answer to 2 significant figures.

.... Pascals

(Total for Question 14 is 4 marks)

Write your name here

Surname

Other names

**In the style of:
Pearson Edexcel
Level 1/Level 2 GCSE (9 - 1)**

Centre Number

Candidate Number

Mathematics

Histograms

Higher Tier

GCSE style questions arranged by topic

Paper Reference
1MA1/1H

You must have: Ruler graduated in centimetres and millimetres, protractor, pair of compasses, pen, HB pencil, eraser.

Total Marks

Instructions

- Use **black** ink or ball-point pen.
- **Fill in the boxes** at the top of this page with your name, centre number and candidate number.
- Answer **all** questions.
- Answer the questions in the spaces provided
  – there may be more space than you need.
- **Calculators may not be used.**
- Diagrams are **NOT** accurately drawn, unless otherwise indicated.
- You must **show all your working out**.

Information

- The total mark for this paper is
- The marks for **each** question are shown in brackets
  – use this as a guide as to how much time to spend on each question.

Advice

- Read each question carefully before you start to answer it.
- Keep an eye on the time.
- Try to answer every question.
- Check your answers if you have time at the end.

Turn over ▶

© Peter Bland

1 The table gives some information about the speeds, in km/h, of 100 cars.

Speed (s km/h)	Frequency
$60 < s \leq 65$	15
$65 < s \leq 70$	25
$70 < s \leq 80$	36
$80 < s \leq 100$	24

$fd = \dfrac{15}{5} = 3$

$fd = \dfrac{25}{5} = 5$

$fd = \dfrac{36}{10} = 3.6$

$fd = \dfrac{24}{20} = 1.2$

(a) On the grid, draw a histogram for the information in the table.

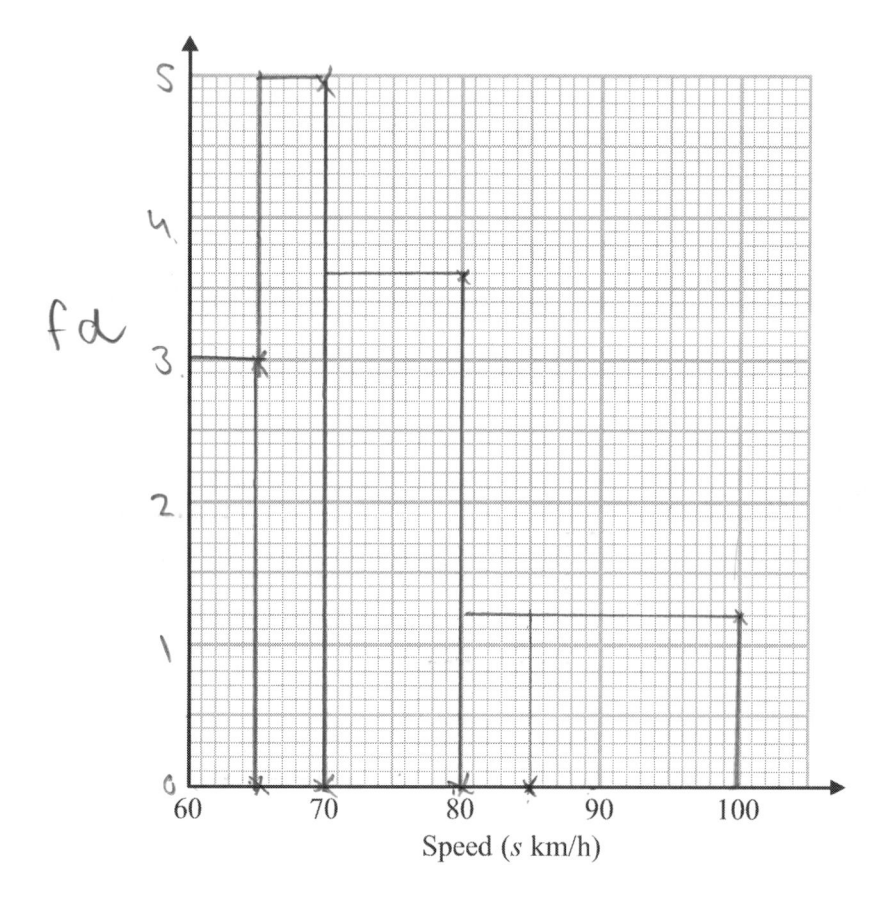

(3)

(b) Work out an estimate for the number of cars with a speed of more than 85 km/h.

$f = fd \times w$
$f = 1.2 \times 15$
$= 18$

................18.................

(2)

(Total for Question 1 is 5 marks)

2. The table gives information about the heights, h centimetres, of plants in a greenhouse.

$fd = \frac{7}{2} = 3.5$
$fd = \frac{14}{2} = 7$
$fd = \frac{16}{4} = 4$
$fd = \frac{22}{8} = 2.75$
$fd = \frac{12}{4} = 3$

Height (h centimetres)	Frequency
$0 < h \leq 2$	7
$2 < h \leq 4$	14
$4 < h \leq 8$	16
$8 < h \leq 16$	22
$16 < h \leq 20$	12

Draw a histogram to show this information.

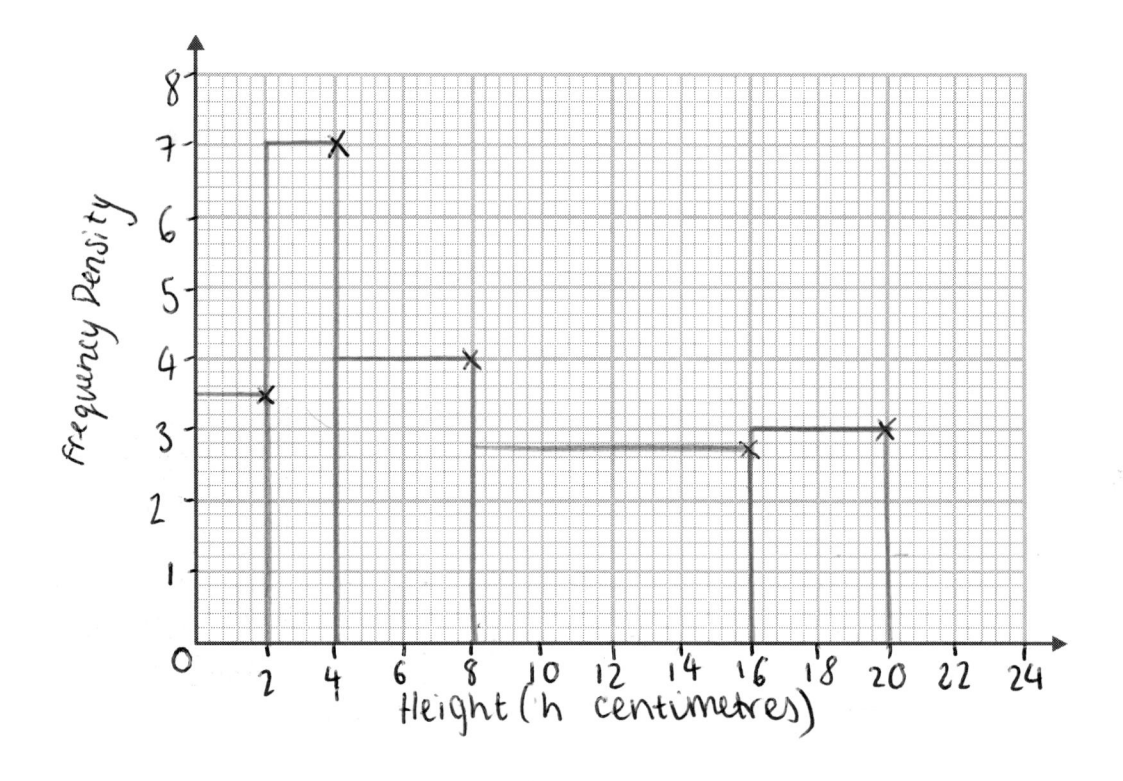

(Total for Question 2 is 3 marks)

3 The table gives information about the ages of the population of a city.

Age (a years)	Number (thousands)
$0 \leq a < 10$	9
$10 \leq a < 20$	8
$20 \leq a < 35$	10
$35 \leq a < 50$	19
$50 \leq a < 55$	4
$55 \leq a < 65$	7
$65 \leq a < 80$	4
$80 \leq a < 100$	1

$fd = \dfrac{9000}{10} = 900$
$fd = \dfrac{8000}{10} = 800$
$fd = \dfrac{10000}{5} = 2000$
$fd = \dfrac{19000}{15} = 1266.7$
$fd = \dfrac{4000}{5} = 800$
$fd = \dfrac{7000}{10} = 700$
$fd = \dfrac{4000}{15} = 266.7$
$fd = \dfrac{1000}{20} = 50$

(a) On the graph paper below, using a scale of 1 cm to represent 10 years on the Age axis, draw a histogram to represent this information.

(4)

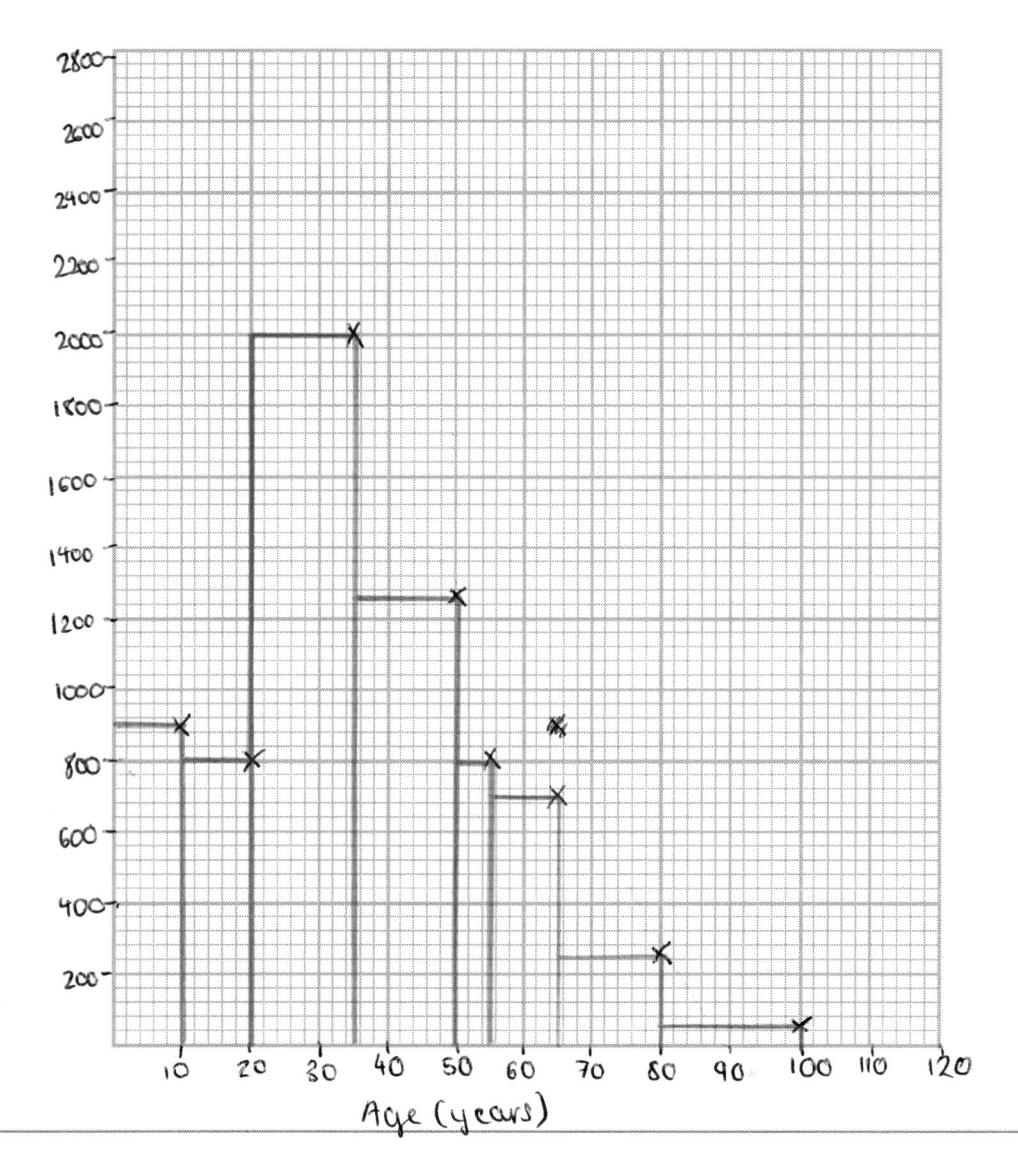

(b) Write down the class interval in which the median lies.

(1)

(c) Calculate, giving your answer in years and months, an estimate of the mean age of the population.

(4)

(Total for Question 3 is 9 marks)

4 A pub has 64 customers one evening.
The table gives information about the lengths, in minutes, of the time the customers stayed for.

Length (x) minutes	Frequency
$0 < x \leqslant 5$	1
$5 < x \leqslant 15$	10
$15 < x \leqslant 30$	17
$30 < x \leqslant 40$	21
$40 < x \leqslant 45$	15

$fd = \frac{1}{5} = 0.2$

$fd = \frac{10}{5} = 2$

$fd = \frac{17}{15} = 1\frac{2}{15}$

$fd = \frac{21}{10} = 2.1$

$fd = \frac{15}{5} = 3$

Draw a histogram for this information.

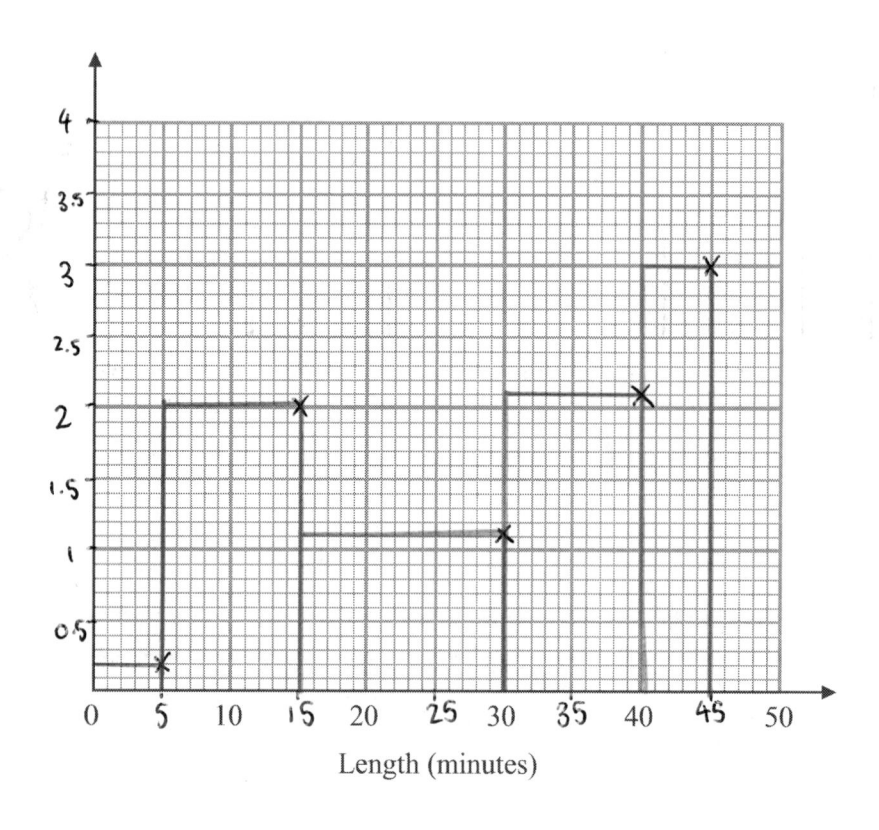

(Total for Question 4 is 4 marks)

5 The incomplete histogram and table show information about the weights of some vehicles.

Weight (w) in kg	Frequency
$0 < w \leqslant 1000$	16
$1000 < w \leqslant 2000$	
$2000 < w \leqslant 4000$	
$4000 < w \leqslant 6000$	14
$6000 < w \leqslant 8000$	
$8000 < w \leqslant 12000$	4

(a) Use the information in the histogram to complete the table.

(2)

(b) Use the information in the table to complete the histogram.

(2)

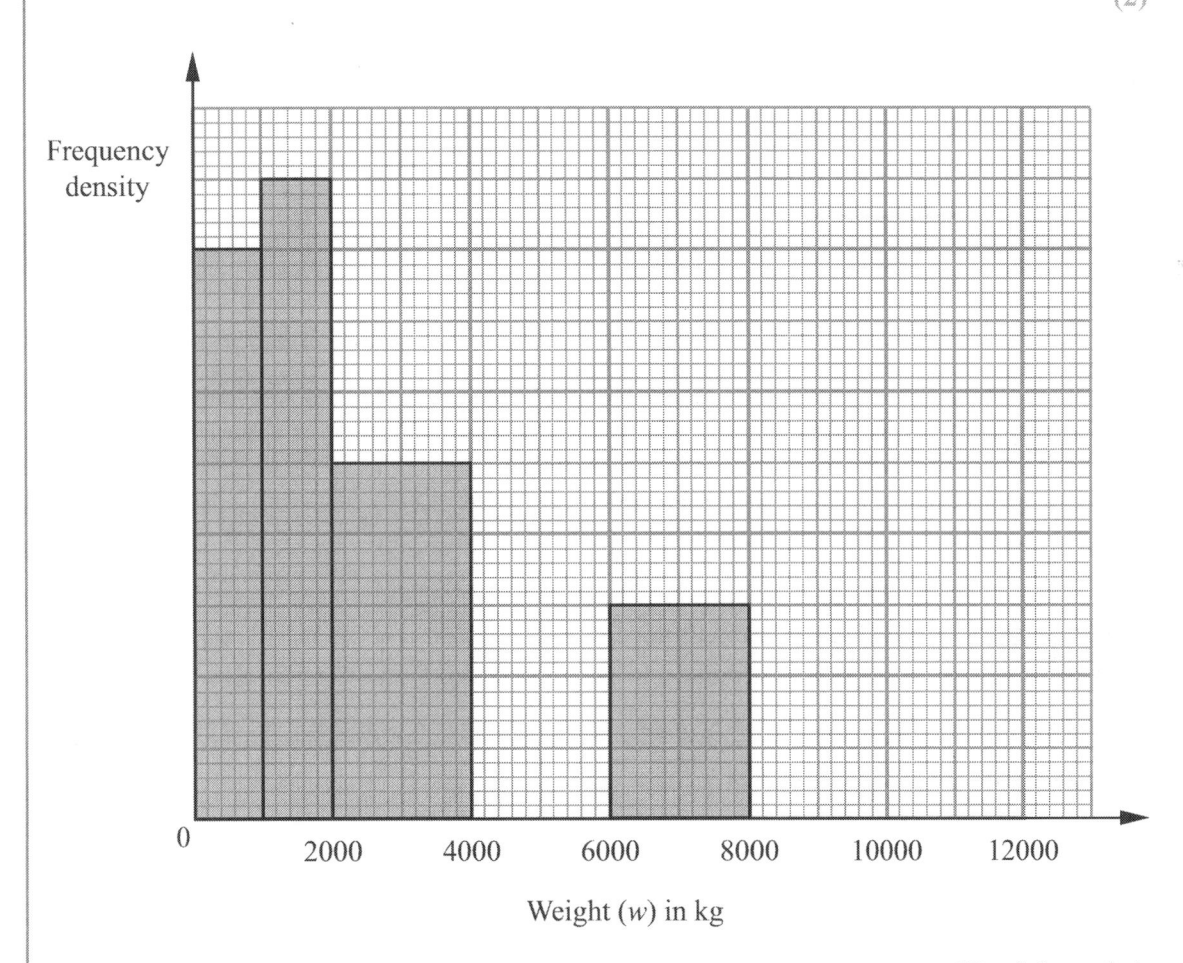

(Total 4 marks)

6. One hundred hikers went for a walk. The times taken by the hikers to complete the walk are summarised in the table.

Time (t)	Number of hikers
$0 \leqslant t < 25$	15
$25 \leqslant t < 35$	11
$35 \leqslant t < 40$	27
$40 \leqslant t < 60$	15
$60 \leqslant t < 90$	15
$90 \leqslant t < 100$	12

(a) Use the information given in the table to calculate an estimate for the mean time taken, to one decimal place.

(3)

(b) Given that the frequency density for the $40 \leqslant t < 60$ time interval is 0.75, complete the histogram to represent this information on the graph paper.

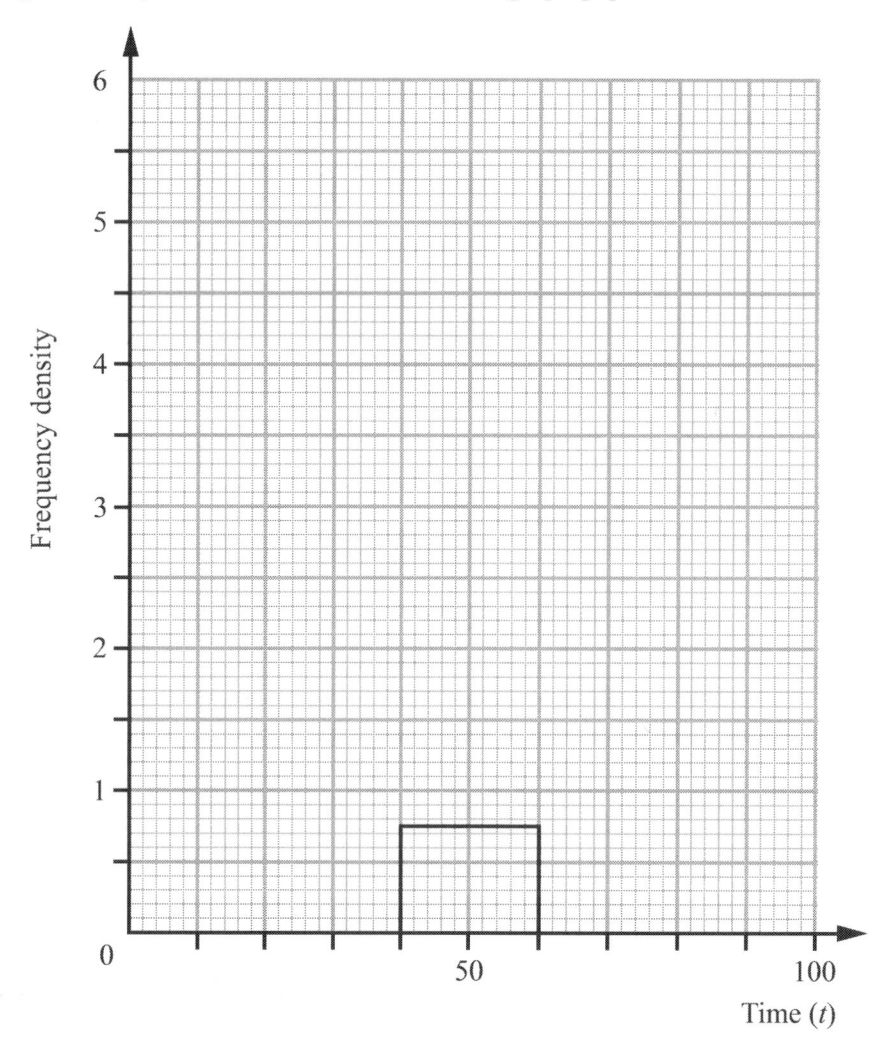

(4)

(Total for Question 6 is 7 marks)

7 The incomplete histogram and table give some information about the distances some cyclists travel each day.

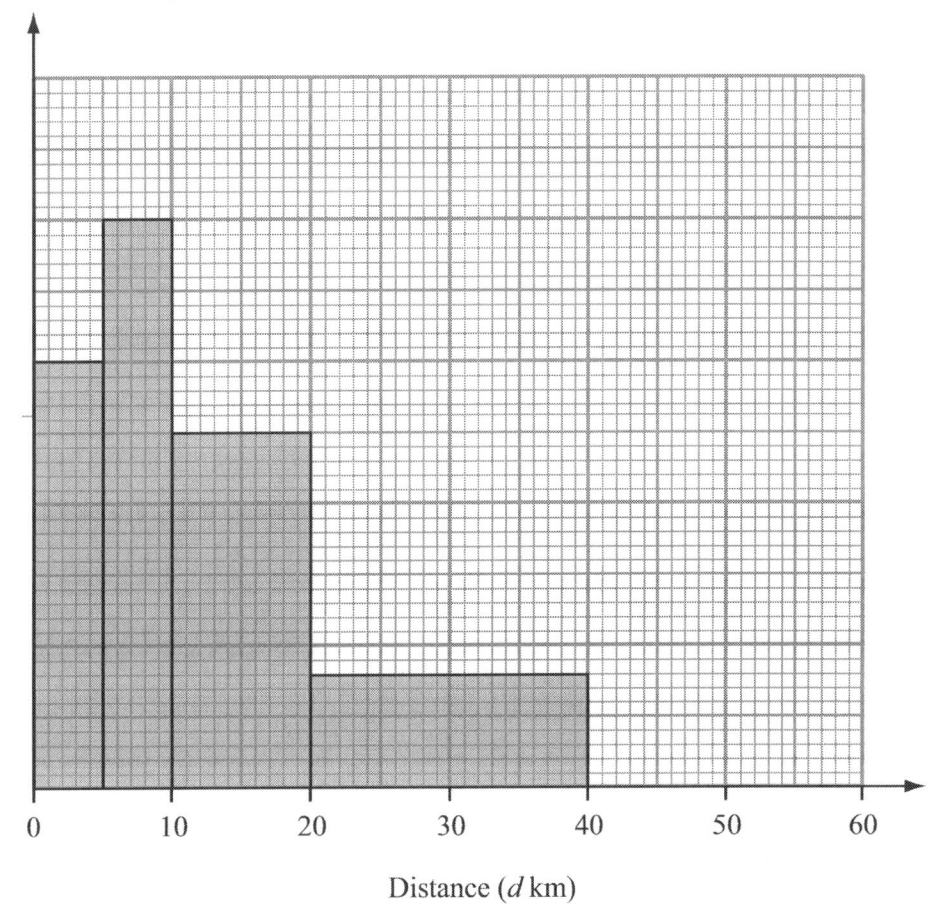

(a) Use the information in the histogram to complete the frequency table.

Distance (d km)	Frequency
$0 < d \leqslant 5$	15
$5 < d \leqslant 10$	20
$10 < d \leqslant 20$	
$20 < d \leqslant 40$	
$40 < d \leqslant 60$	15

(2)

(b) Use the information in the table to complete the histogram.

(1)

(Total for Question 7 is 3 marks)

8 Terry asked the students in his class how many hours they played on computers last week.

The incomplete histogram was drawn using his results.

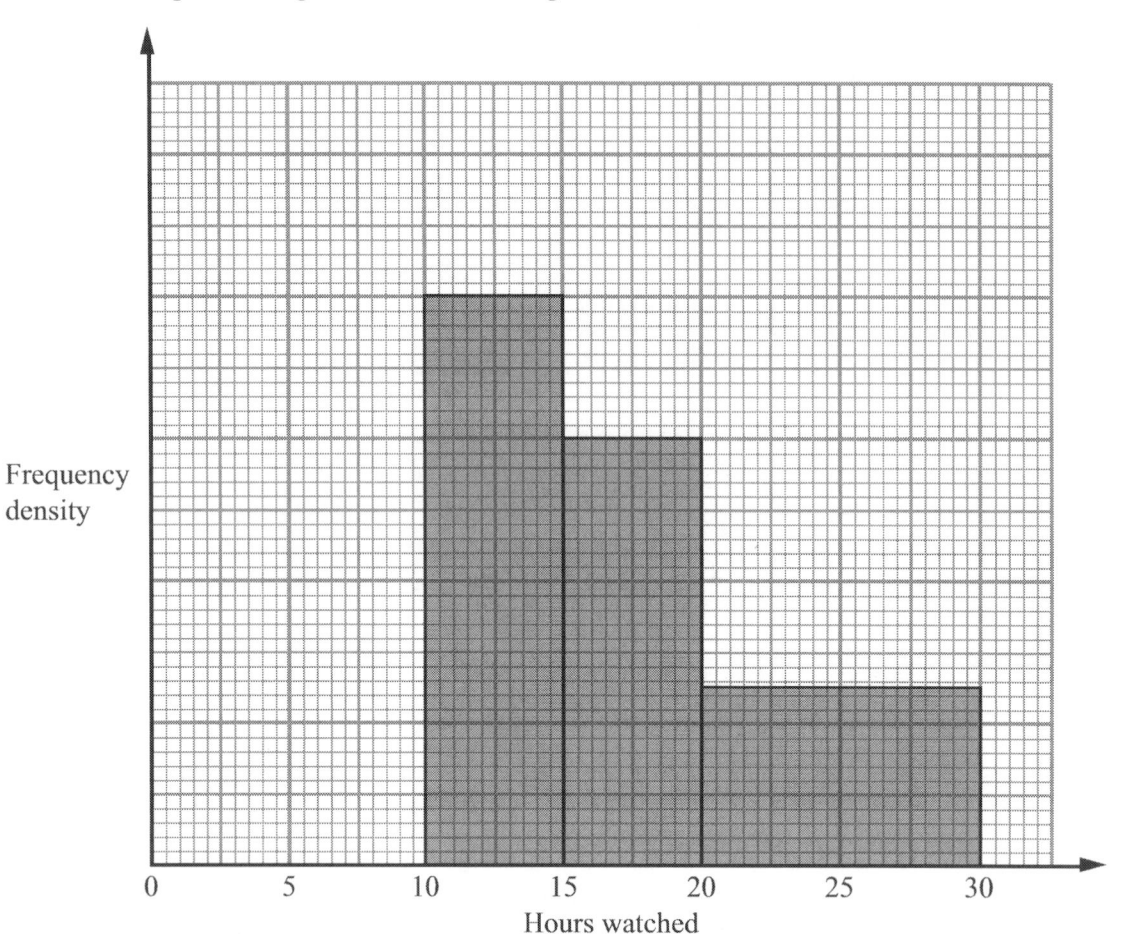

Eight students played for between 10 and 15 hours. Six students played for between 0 and 10 hours.

(a) Use this information to complete the histogram.

(2)

No students watched television for more than 30 hours.

(b) Work out how many students Terry asked.

....

(2)

(Total for Question 8 is 4 marks)

9 Some trains from London to Birmingham were late.
   The incomplete table and histogram gives some information about how late the trains were.

Minutes late (t)	Frequency
$0 < t \leqslant 5$	16
$5 < t \leqslant 10$	10
$10 < t \leqslant 20$	
$20 < t \leqslant 30$	
$30 < t \leqslant 50$	6

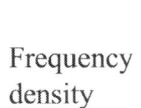

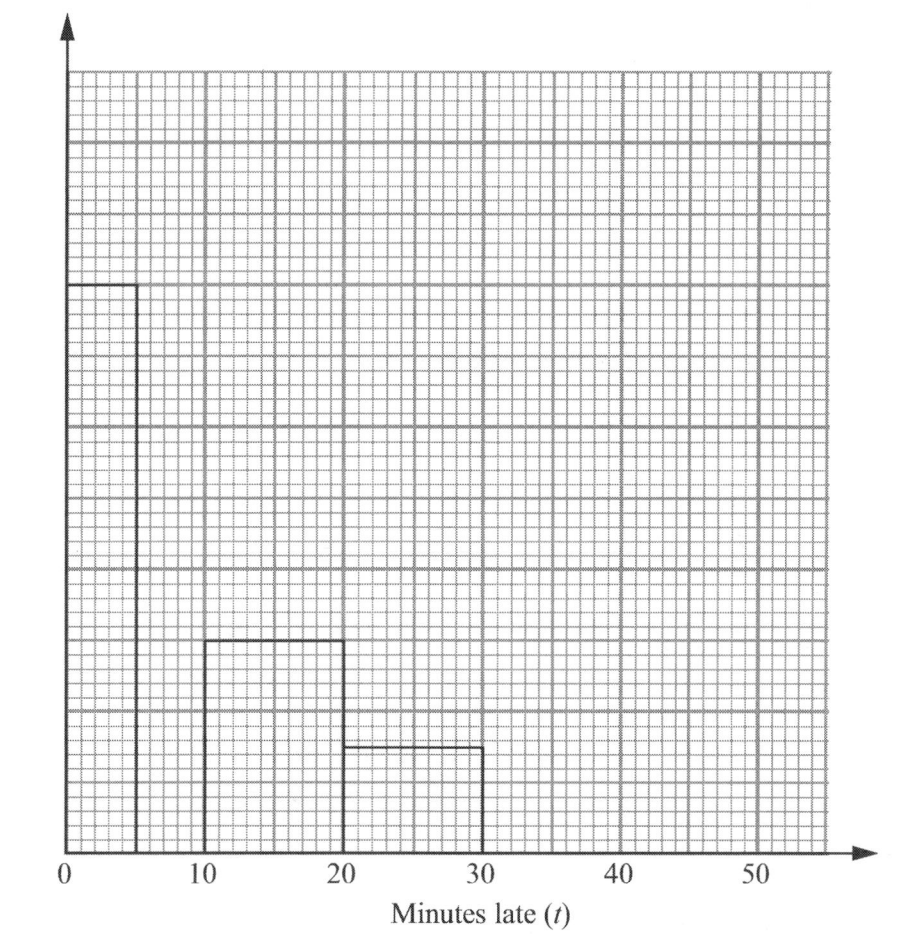

(a) Use the information in the histogram to complete the table.
(2)

(b) Use the information in the table to complete the histogram.
(2)

(Total for Question 9 is 4 marks)

10 The incomplete table and histogram give some information about the heights of some tomato plants in a greenhouse.

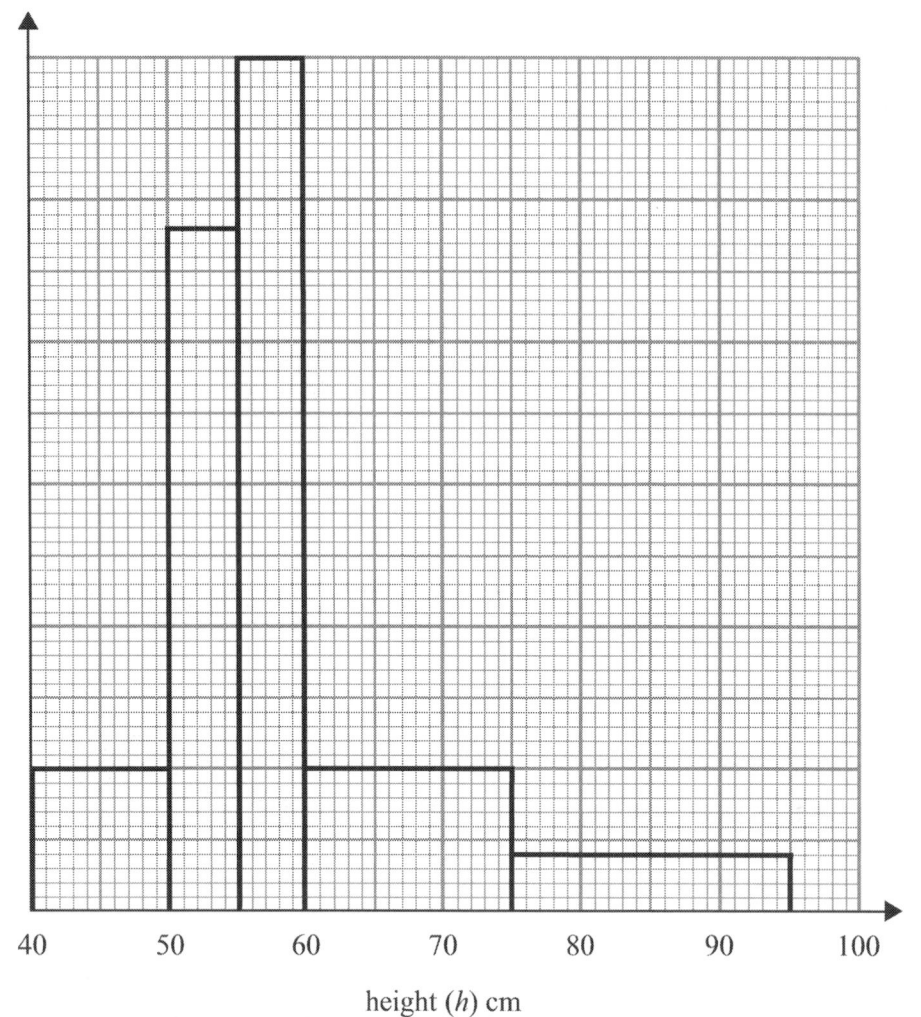

Use the information in the histogram to complete the frequency table.

Height (h) cm	Frequency
$40 \leqslant h < 50$	10
$50 \leqslant h < 55$	
$55 \leqslant h < 60$	
$60 \leqslant h < 75$	15
$75 \leqslant h < 95$	8

(Total for Question 10 is 2 marks)

11 The incomplete table and histogram give some information about the weights (in kg) of some boxes.

Weight (w kg)	Frequency
$100 < w \leqslant 130$	30
$130 < w \leqslant 150$	
$150 < w \leqslant 160$	
$160 < w \leqslant 180$	40
$180 < w \leqslant 210$	18

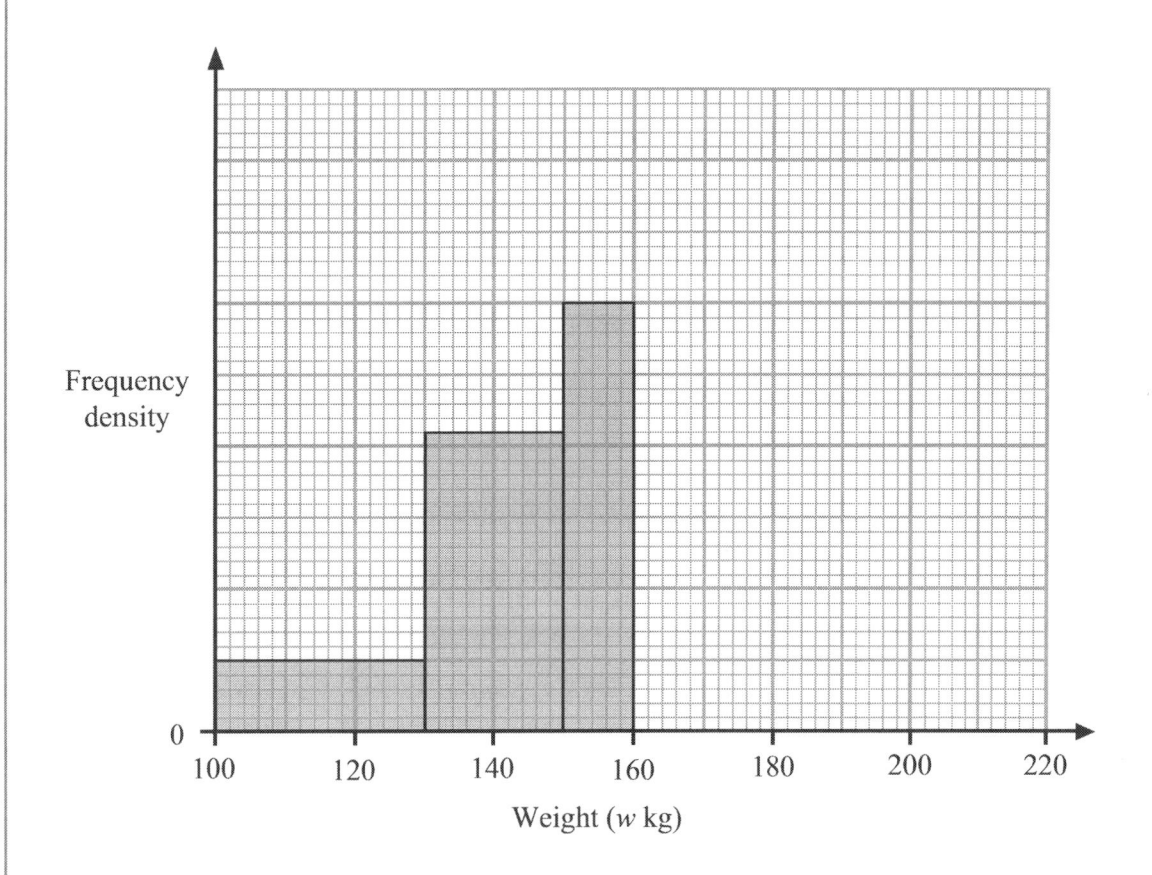

(a) Use the histogram to complete the table.

(2)

(b) Use the table to complete the histogram.

(2)

(Total for Question 11 is 4 marks)

12 The table and histogram show information about the length of time it took 165 adults to drink some water.

Time (t seconds)	Frequency
$0 < t \leqslant 10$	20
$10 < t \leqslant 15$	
$15 < t \leqslant 17.5$	30
$17.5 < t \leqslant 20$	40
$20 < t \leqslant 25$	
$25 < t \leqslant 40$	

None of the adults took more than 40 seconds to drink the water

(a) Use the table to complete the histogram.

(2)

(b) Use the histogram to complete the table.

(2)

2010

Frequency density

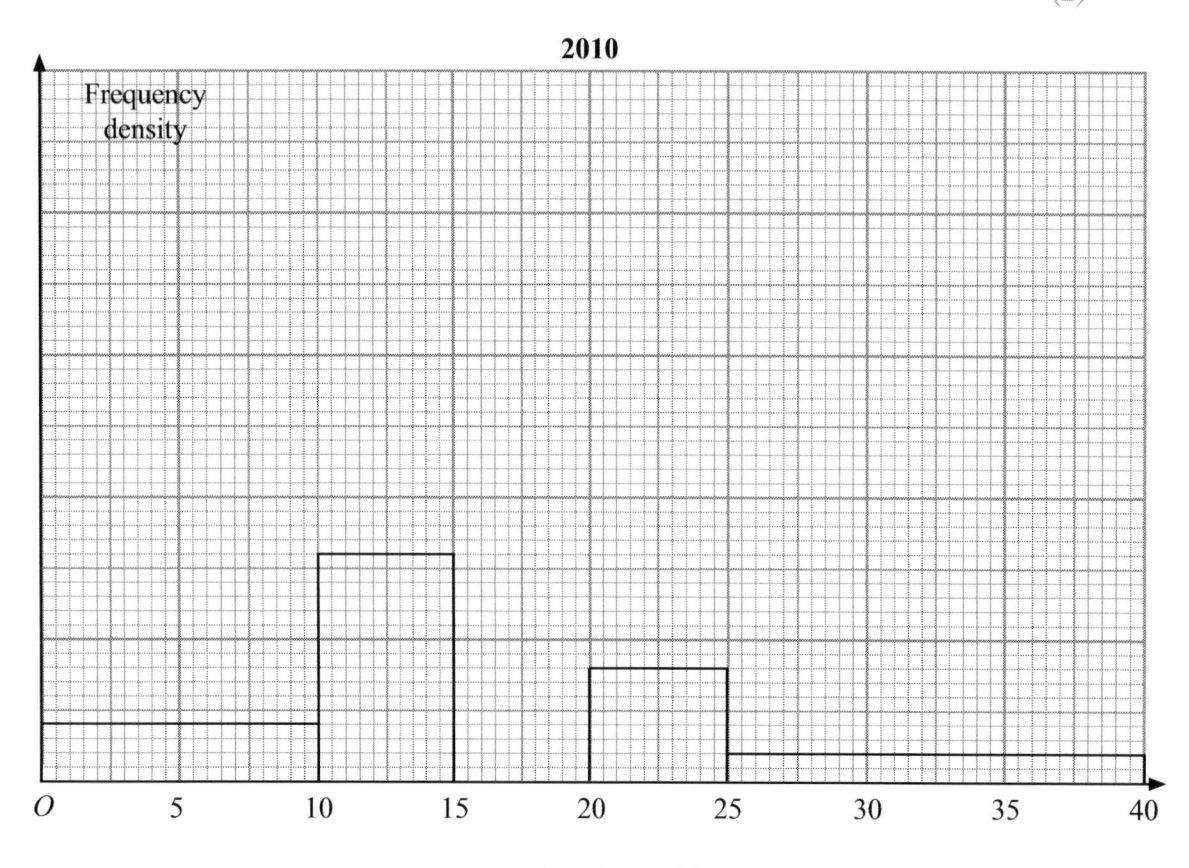

Time (seconds)

2011

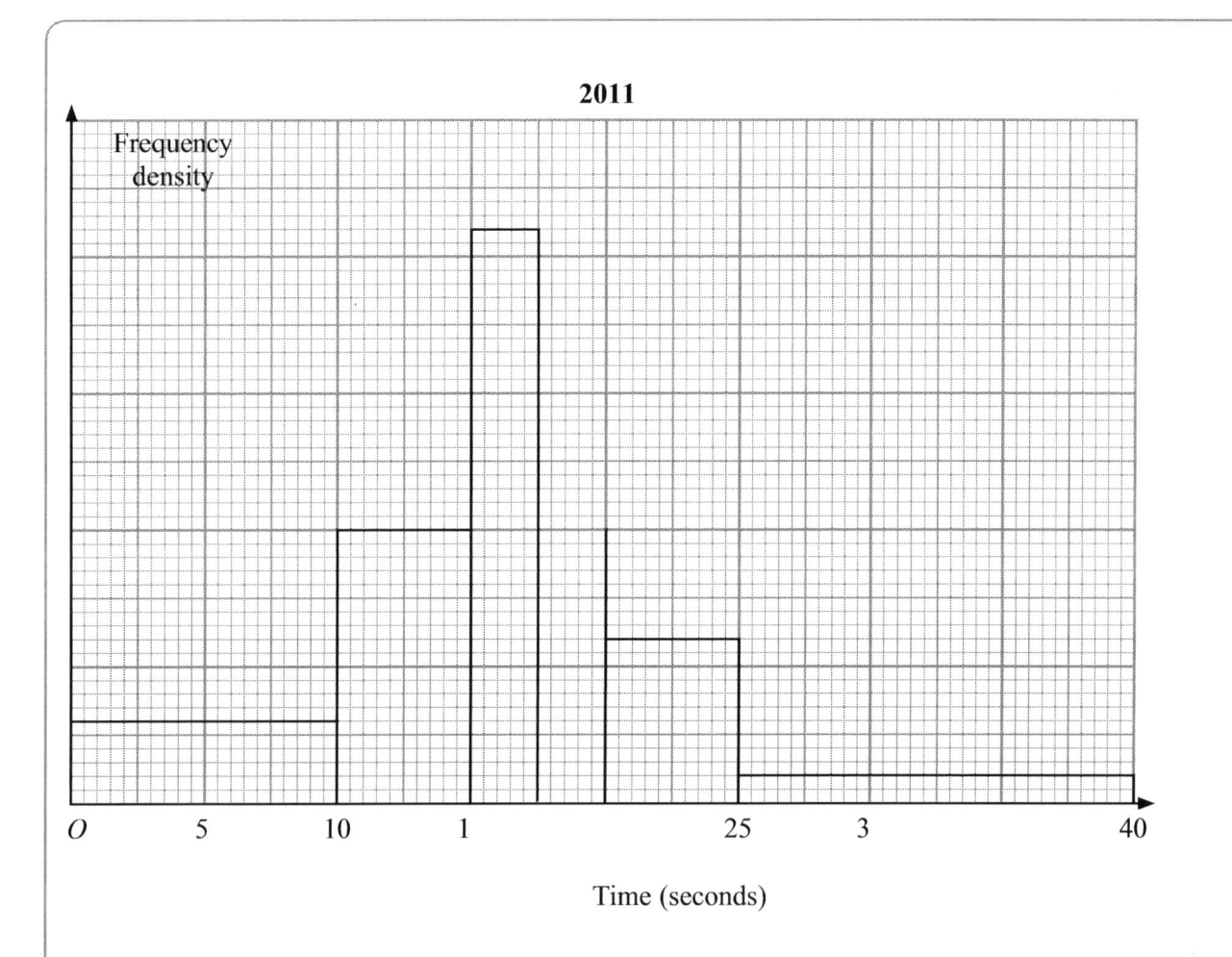

[Histogram: Frequency density vs Time (seconds), with x-axis marks at 5, 10, 1(5), 25, 3(0), 40]

The histogram shows information about the time it took some children to drink the water. None of the children took more than 40 seconds to drink the water.

110 children took up to 12.5 seconds to drink the water.

(c) Work out an estimate for the number of children who took 21 seconds or more to drink the water.

....
(3)

(Total for Question 12 is 7 marks)

13 David recorded the lengths of time, in hours, that some adults watched TV last week.

The table shows information about his results.

Length of time (t hours)	Frequency
$0 \leqslant t < 10$	6
$10 \leqslant t < 15$	8
$15 \leqslant t < 20$	15
$20 \leqslant t < 40$	5

David made some mistakes when he drew a histogram for this information.

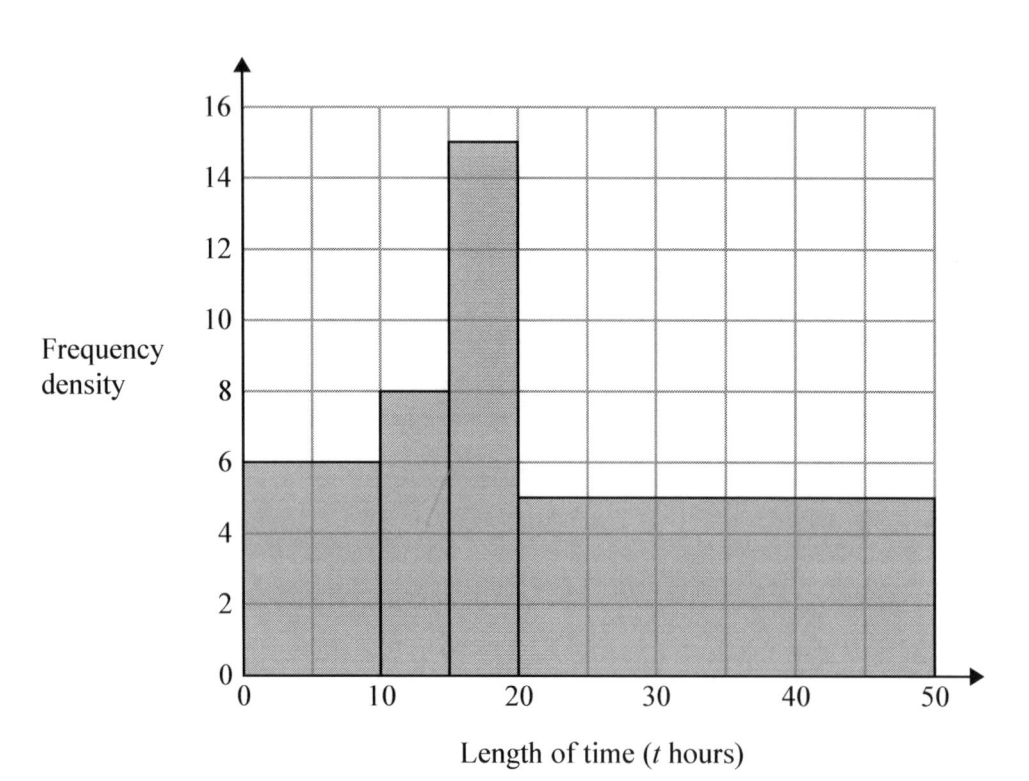

Write down **two** mistakes David made.

1 ....

....

2 ....

....

(Total for Question 13 is 2 marks)

14 The histogram shows the ages, in years, of members of a chess club.

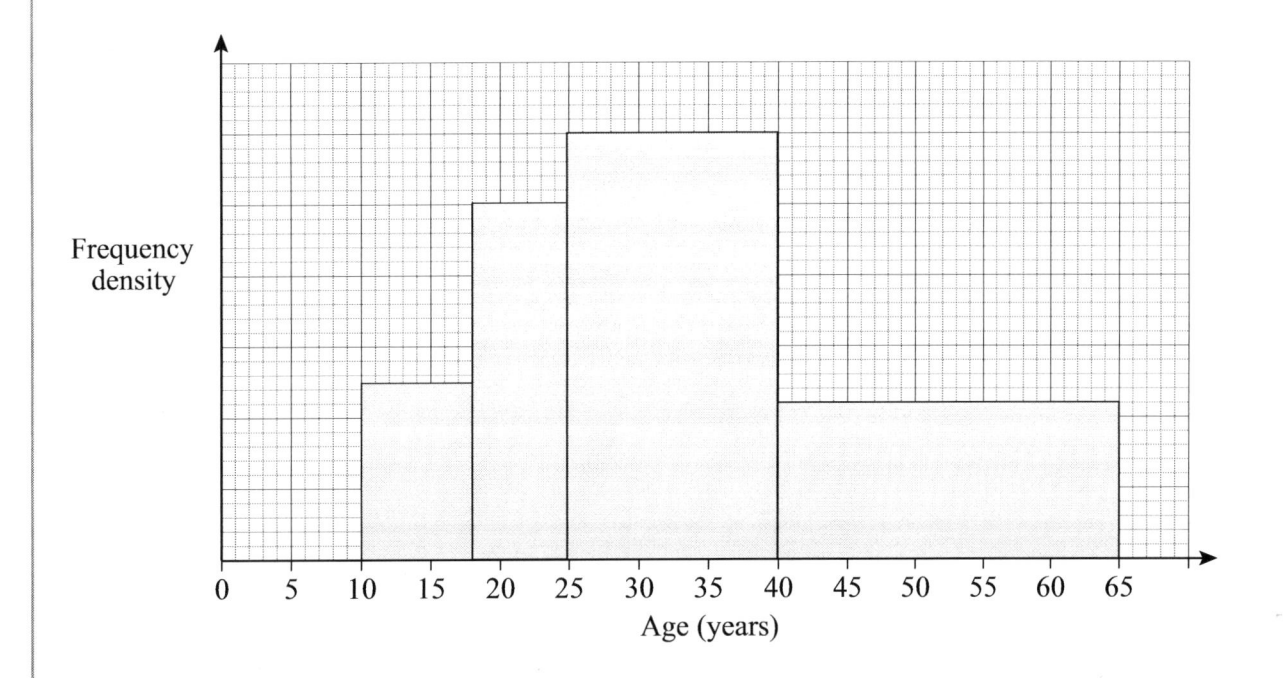

There are 22 members with ages in the range $40 \leqslant \text{age} < 65$

Work out the number of members with ages in the range $25 \leqslant \text{age} < 40$

....
(4)

(Total for Question 14 is 4 marks)

15 Joe works for a computer service centre.

The table shows some information about the length of time, t minutes, of the phone calls Joe had.

Time (t minutes)	$0 < t \leq 10$	$10 < t \leq 15$	$15 < t \leq 20$	$20 < t \leq 30$	$30 < t \leq 45$
Number of calls	12	15	13	18	3

On the grid, draw a histogram to show this information.

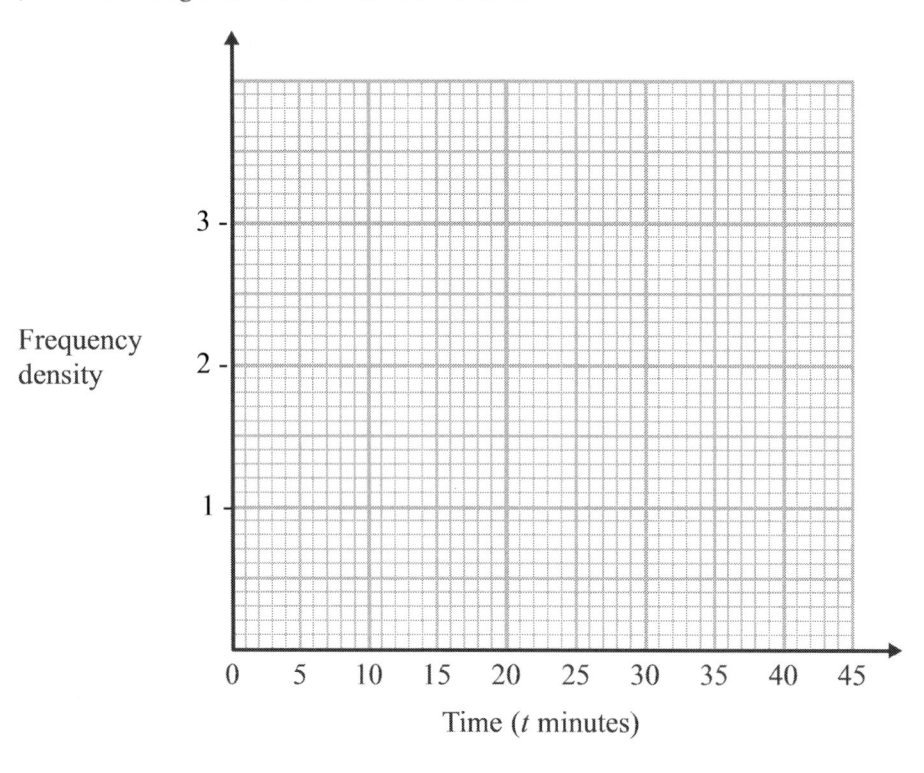

(Total for Question 15 is 3 marks)

Write your name here

Surname

Other names

In the style of:
Pearson Edexcel
Level 1/Level 2 GCSE (9 - 1)

Centre Number

Candidate Number

Mathematics
Locus and Constructions
Higher Tier

GCSE style questions arranged by topic

Paper Reference
1MA1/1H

You must have: Ruler graduated in centimetres and millimetres, protractor, pair of compasses, pen, HB pencil, eraser.

Total Marks

Instructions

- Use **black** ink or ball-point pen.
- **Fill in the boxes** at the top of this page with your name, centre number and candidate number.
- Answer **all** questions.
- Answer the questions in the spaces provided
  – *there may be more space than you need.*
- **Calculators may not be used.**
- Diagrams are **NOT** accurately drawn, unless otherwise indicated.
- You must **show all your working out**.

Information

- The total mark for this paper is
- The marks for **each** question are shown in brackets
  – *use this as a guide as to how much time to spend on each question.*

Advice

- Read each question carefully before you start to answer it.
- Keep an eye on the time.
- Try to answer every question.
- Check your answers if you have time at the end.

Turn over ▶

© Peter Bland

1 (a) Draw the locus of all points which are equidistant from the points C and D.

$C \times$ $\times D$

(2)

(b) Draw the locus of all points that are exactly 3 cm from the line EF.

E F

(2)

(Total for Question 1 is 4 marks)

2 Draw the locus of all points which are equidistant from the lines *XY* **and** *XZ*.

Y
|
|
|
|
X_____Z

(Total for Question 2 is 2 marks)

3 The map shows part of a golf course.

A golfer has to hit a ball so that its path between AB and CD is a straight line and is always the same distance from *A* as from *B*

On the map, draw the path the ball should take.

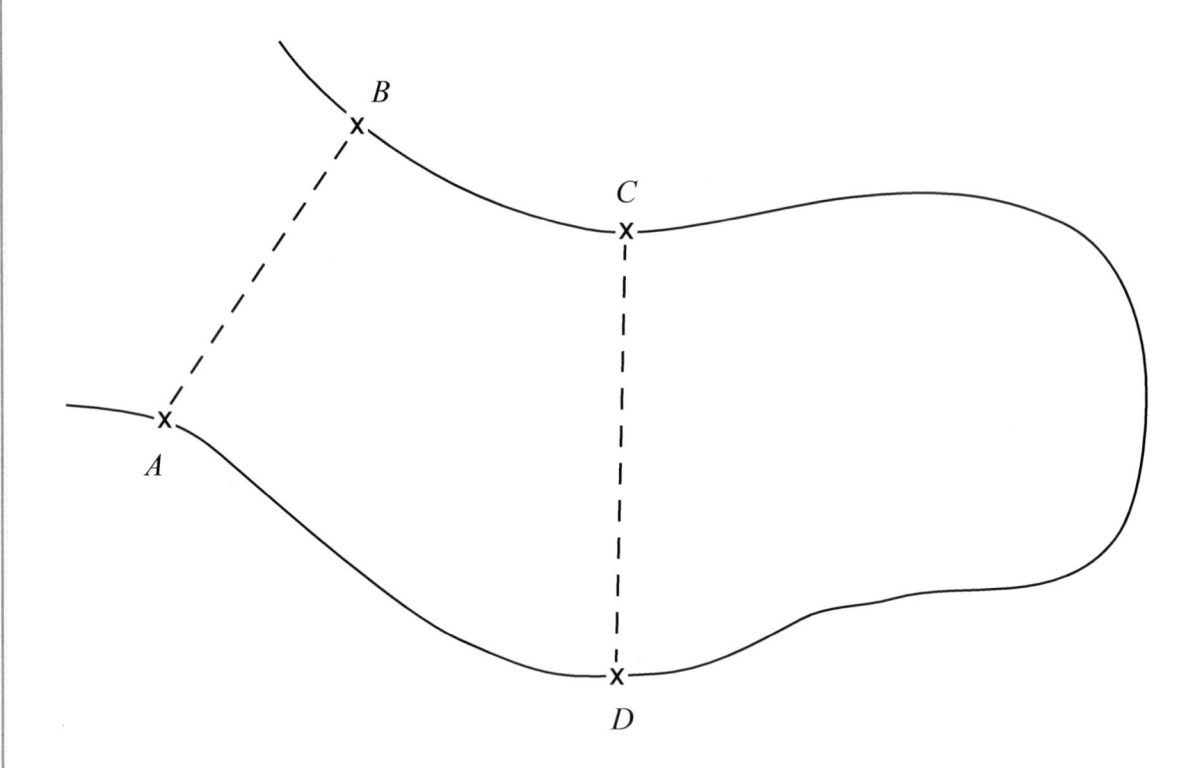

(Total for Question 3 is 2 marks)

4

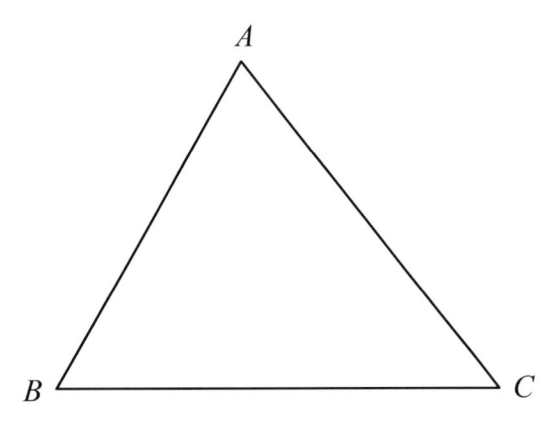

ABC is a triangle.

Shade the region inside the triangle which is **both**

        less than 4 centimetres from the point B

and   closer to the line AC than the line AB.

(Total for Question 4 is 4 marks)

5 Here is a map.
   The map shows two towns, Knutsford and Macclesfield.

Knutsford
×

×
Macclesfield

Scale: 1 cm represents 10 km

A company is going to build a glasshouse.

The glasshouse will be less than 30 km from Knutsford **and** less than 50 km from

Macclesfield. Shade the region on the map where the company can build the glasshouse.

(Total for Question 5 is 3 marks)

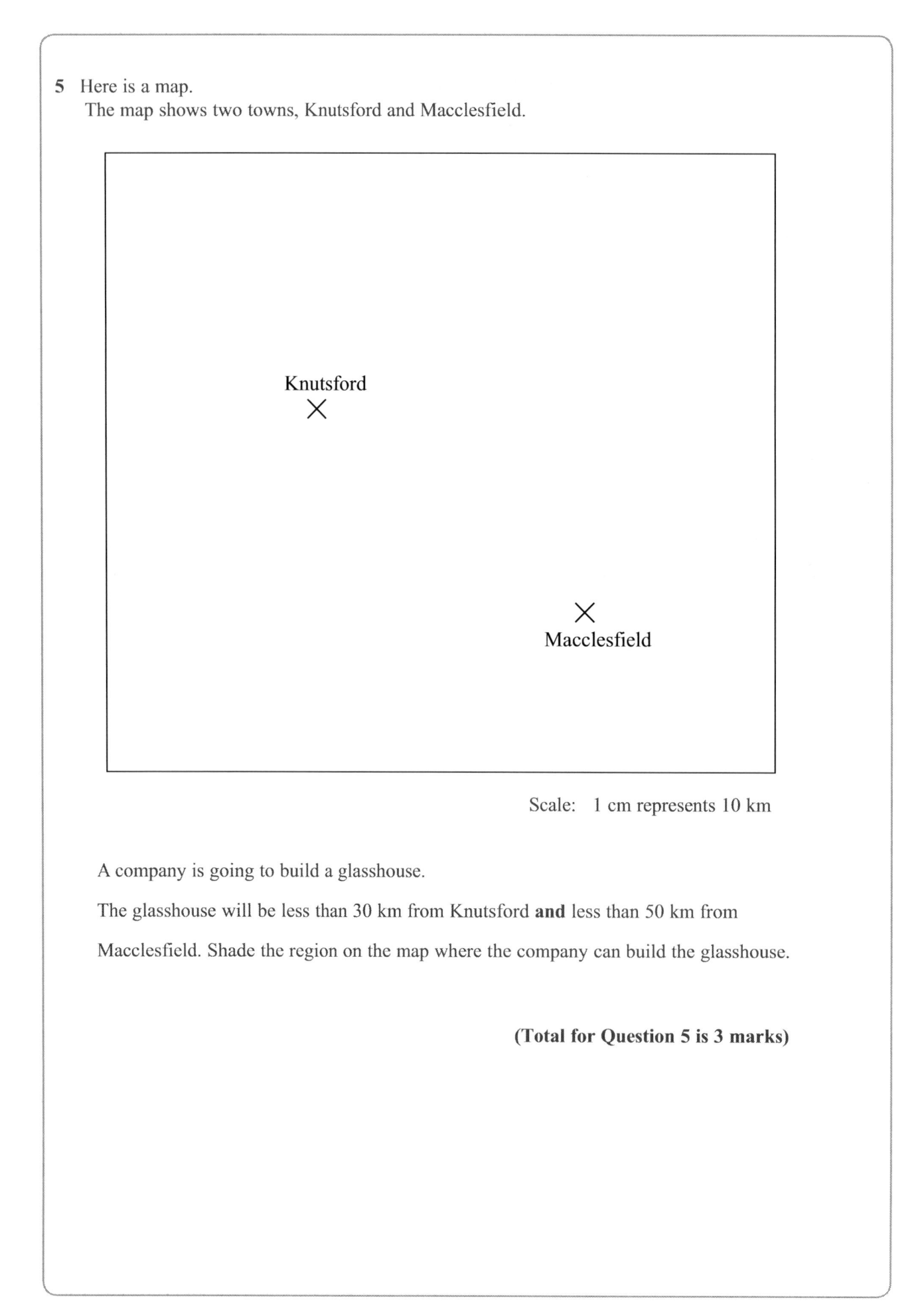

6   The diagram represents a solid made from 5 identical cubes.

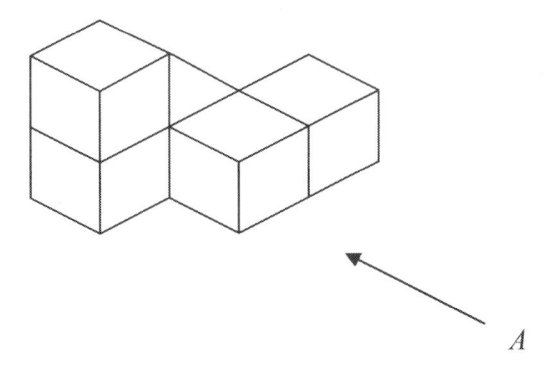

On the grid below, draw the view of the solid from direction *A*.

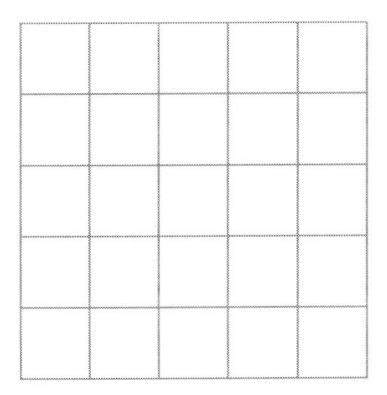

(Total for Question 6 is 2 marks)

7   Here are the plan and front elevation of a solid shape.

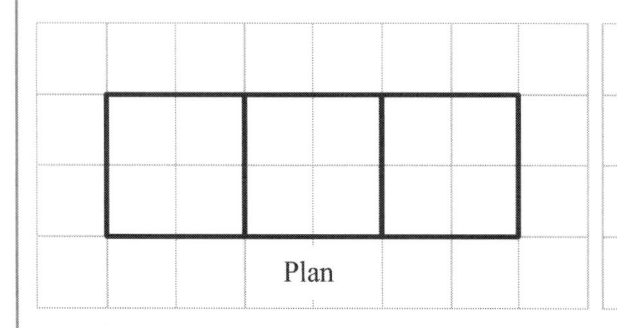

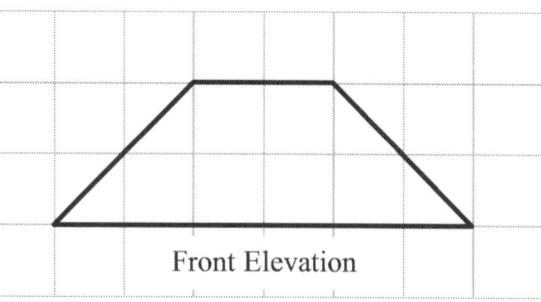

Plan

Front Elevation

(a) On the grid below, draw the side elevation of the solid shape.

(2)

(b) In the space below, draw a sketch of the solid shape.

(2)

(Total for Question 7 is 4 marks)

8 In the space below, use ruler and compasses to **construct** an equilateral triangle with sides of length 6 centimetres.

You must show all your construction lines.

One side of the triangle has already been drawn for you.

(Total for Question 8 is 2 marks)

9 Here is a sketch of a quadrilateral.

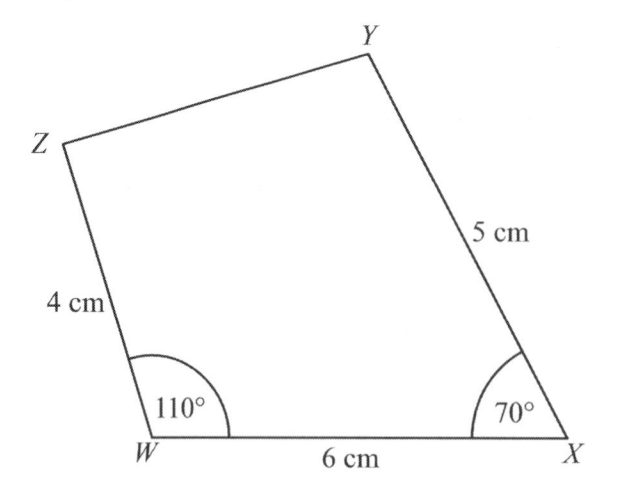

Diagram **NOT** accurately drawn

- 4 cm (ZW)
- 5 cm (YX)
- 6 cm (WX)
- 110° at W
- 70° at X

Make an accurate drawing of the quadrilateral WXYZ in the space below.
The point W, marked with a cross (×), has been drawn for you.

W ×

(Total for Question 9 is 4 marks)

10 (a) On the grid, draw an isosceles triangle.

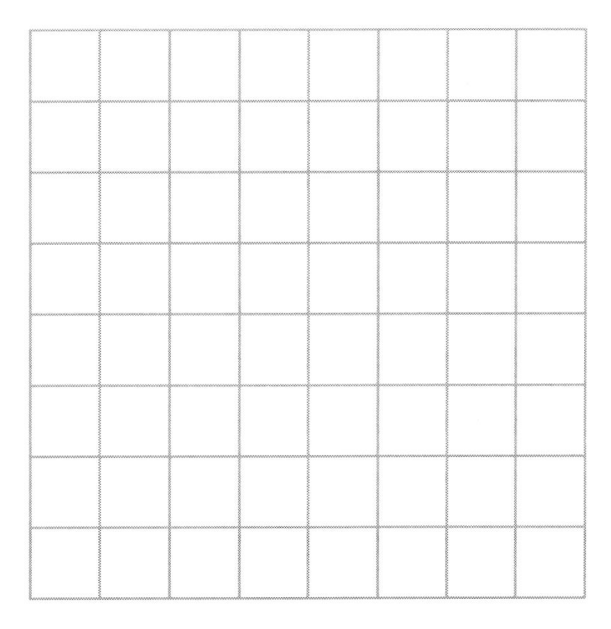

(1)

(b) On the grid, draw a rectangle with an area of 20 cm².

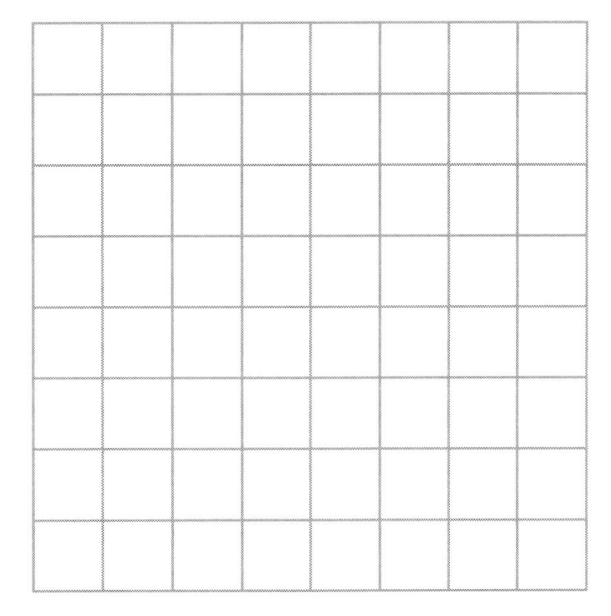

(2)

(Total for Question 10 is 3 marks)

11 (a) Measure the length of the line *AB*.
Give your answer in centimetres.

A ———————————————————B

.... cm
(1)

(b) Measure the size of angle *y*.

.... °
(1)

(c) In the space below, draw accurately a circle of radius 4 cm.
Use the point *C* as the centre of your circle.

×*C*

(1)

(Total for Question 11 is 3 marks)

12 Use ruler and compasses to **construct** the perpendicular bisector of the line *AB*.

You must show all your construction lines.

A ——————————— *B*

(2)

(Total for Question 12 is 2 marks)

13 Use ruler and compasses to **construct** an angle of 30° at *T*.
You **must** show all your construction lines.

T ─────────────────

(Total for Question 13 is 3 marks)

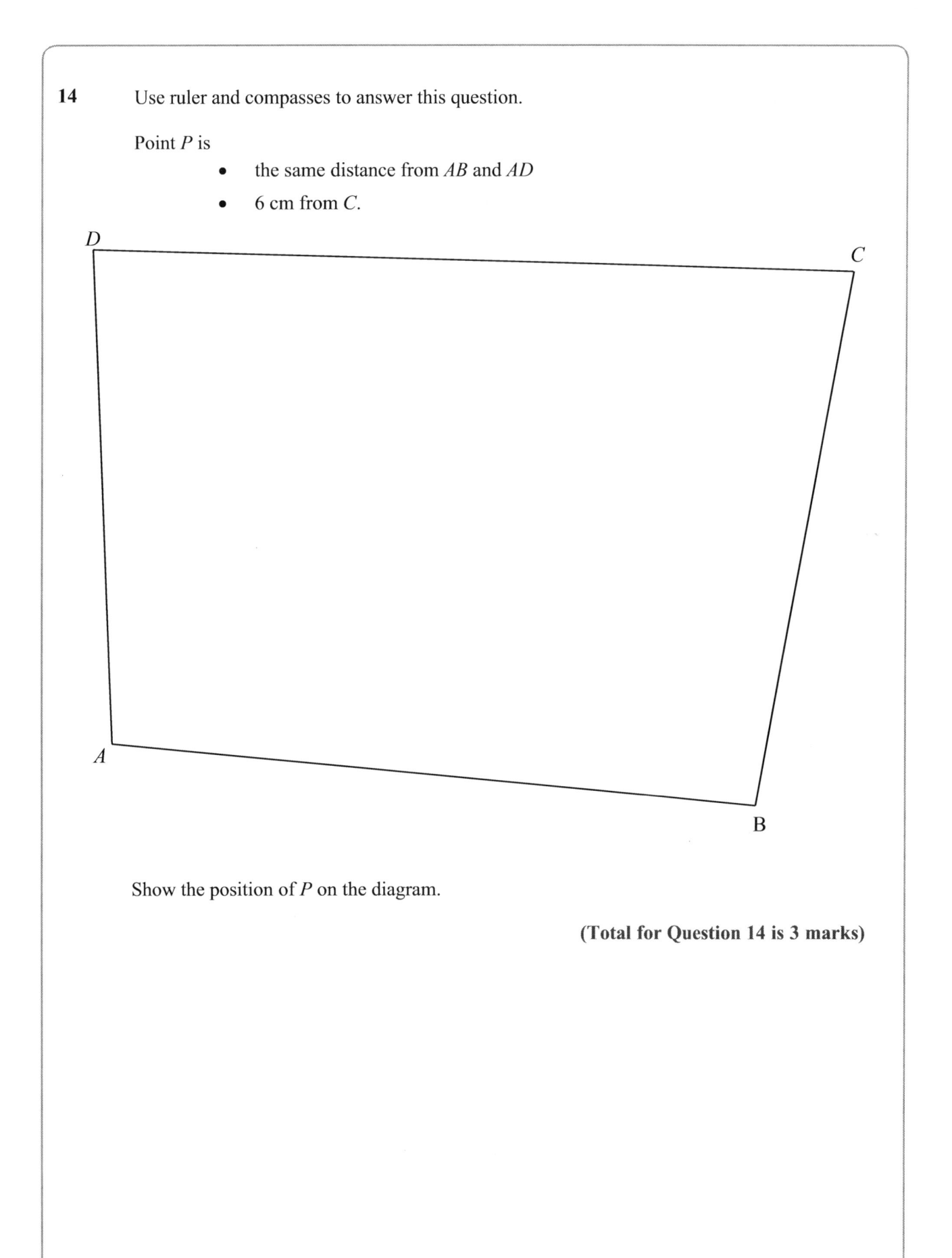

14 Use ruler and compasses to answer this question.

Point P is
- the same distance from AB and AD
- 6 cm from C.

Show the position of P on the diagram.

(Total for Question 14 is 3 marks)

15 Here is a map.
The position of a ship, S, is marked on the map.

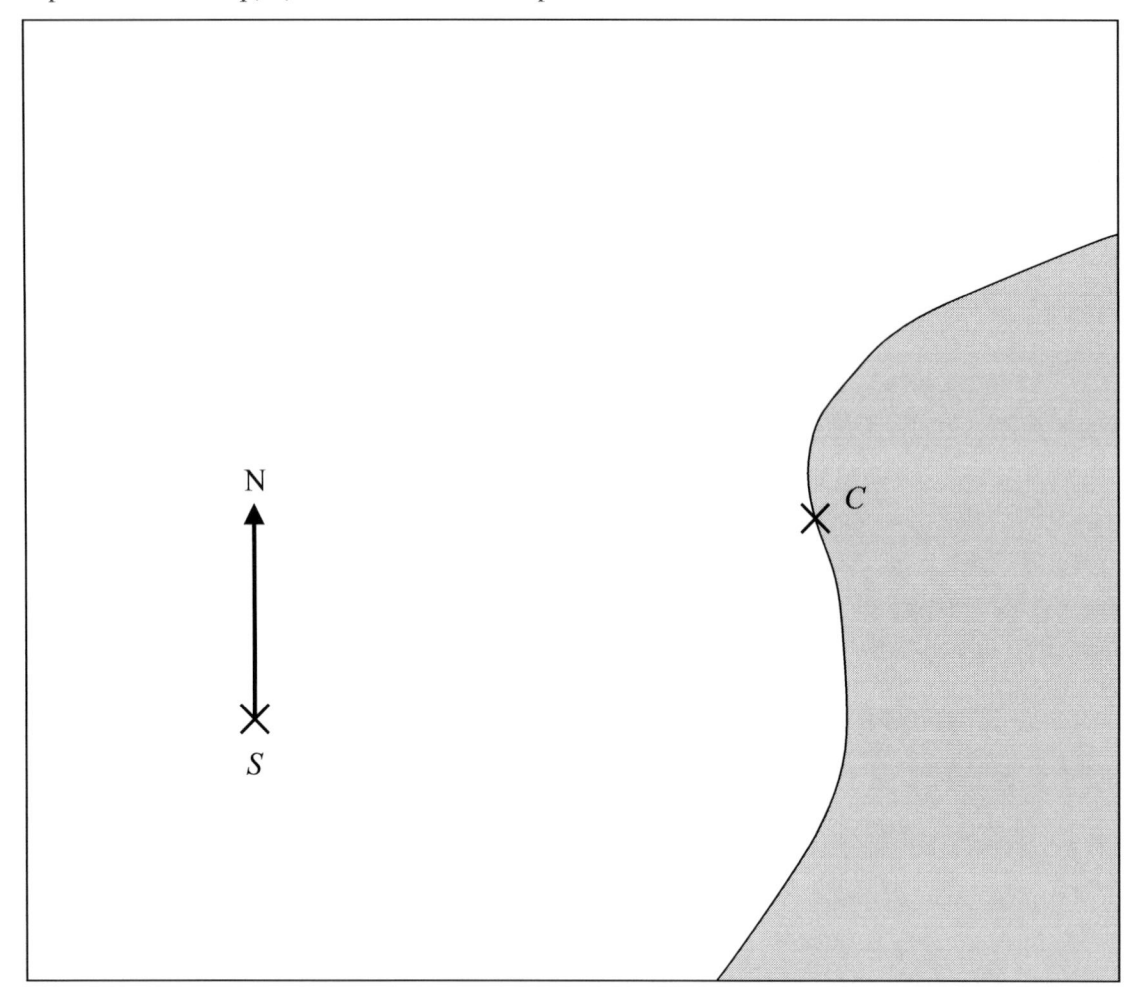

Scale   1 cm represents 100 m

Point C is on the coast.
Ships must not sail closer than 500 m to point C.

The ship sails on a bearing of 037°

Will the ship sail closer than 500 m to point C?
You must explain your answer.

(Total for Question 14 is 3 marks)

16 The diagram shows a garden in the shape of a rectangle.

The scale of the diagram is 1 cm represents 2 m.

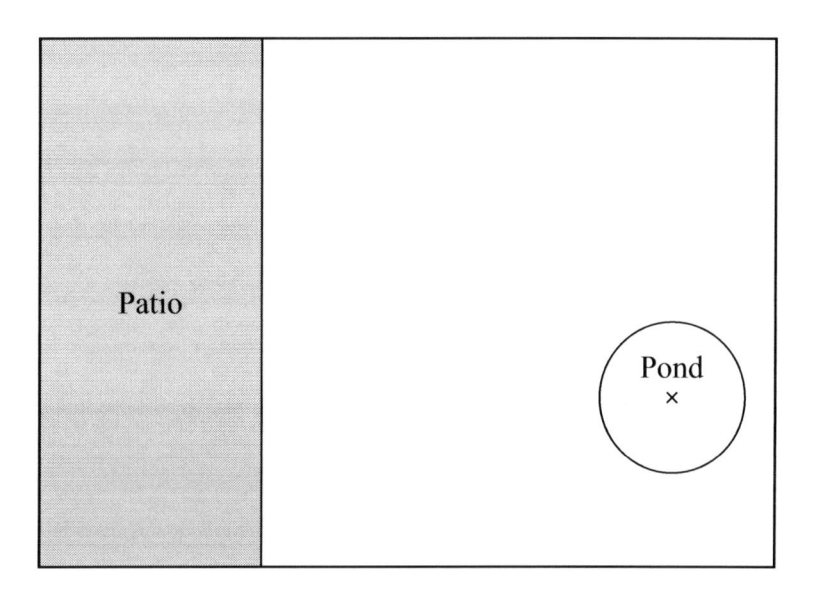

Scale: 1 cm represents 2 m

Dominic is going to plant a tree in the garden.

The tree must be more than 3 metres from the patio
        and more than 6 metres from the centre of the pond.

On the diagram, shade the region where Dominic can plant the tree.

(Total for Question 15 is 3 marks)

17 Here is a scale drawing of a rectangular garden *ABCD*.

A   *B*

D   *C*

Scale: 1 cm represents 1 metre.

Chris wants to plant a tree in the garden

        at least 5 m from point *C*, nearer to *AB* than to *AD*
            and   less than 3 m from *DC*.

On the diagram, shade the region where Chris can plant the tree.

(Total for Question 16 is 4 marks)

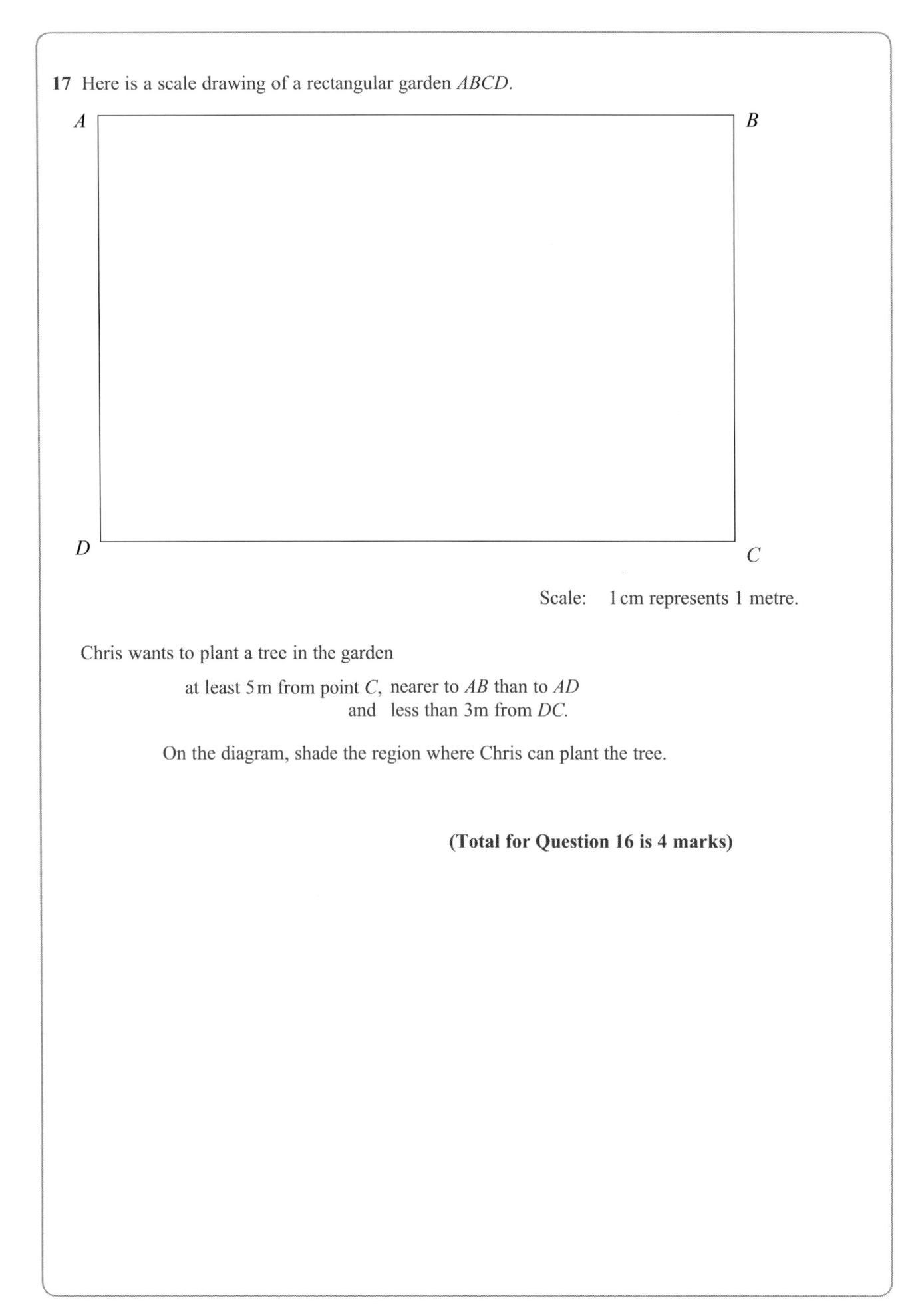

18 The diagram shows the positions of two shops, *A* and *B*, on a map.

× Stockport

×
Altrincham

The scale of the map is 1 cm represents 5 km. Sophie wants to build a warehouse. The warehouse needs to be less than 10 km from Altrincham and less than 20 km from Stockport.

Show by shading where Sophie can build the warehouse.

(Total for Question 9 is 3 marks)

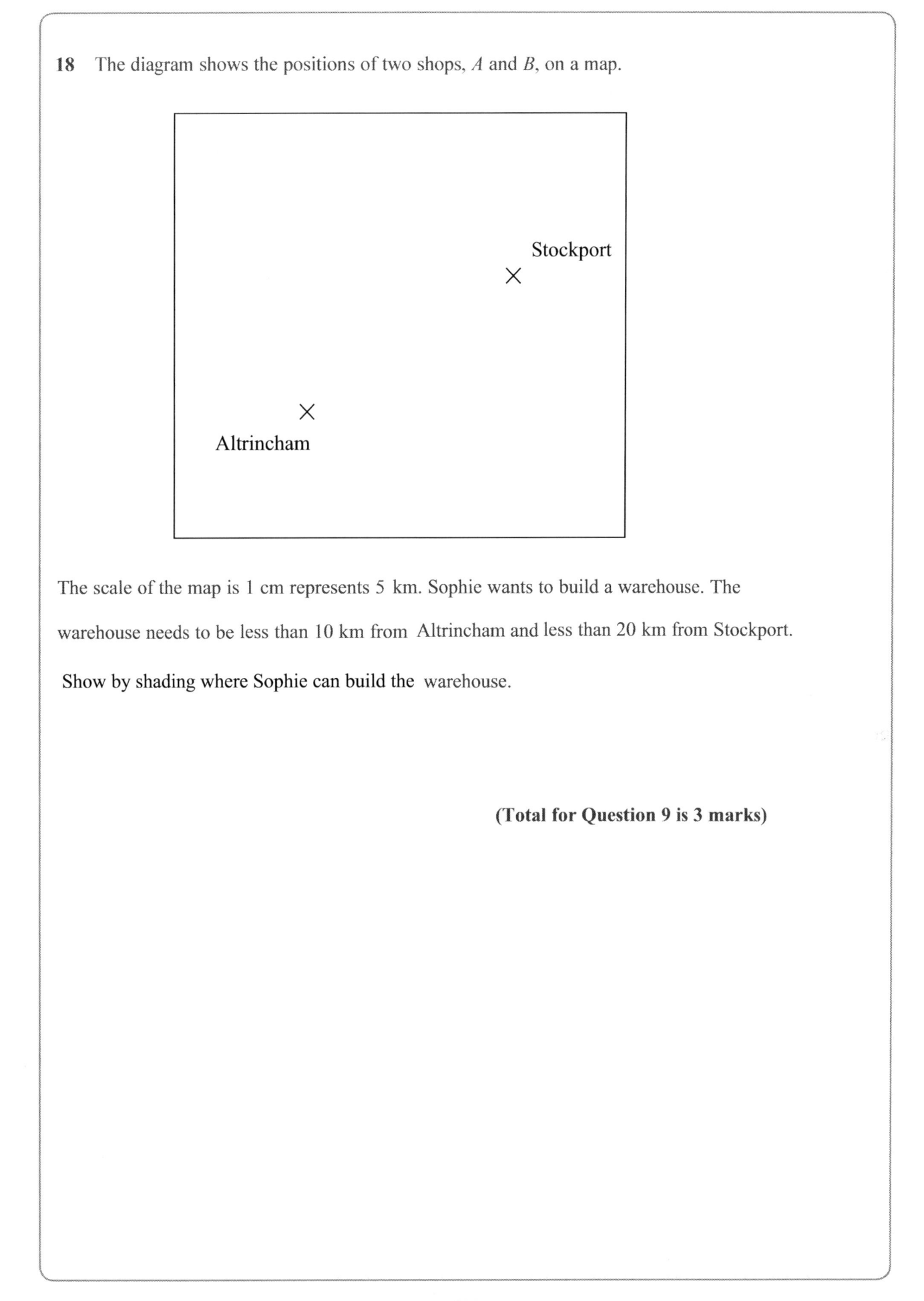

Write your name here

Surname

Other names

In the style of:
Pearson Edexcel
Level 1/Level 2 GCSE (9 - 1)

Centre Number

Candidate Number

Mathematics

Number

Foundation Tier

GCSE style questions arranged by topic

Paper Reference
1MA1/1F

You must have: Ruler graduated in centimetres and millimetres, protractor, pair of compasses, pen, HB pencil, eraser.

Total Marks

Instructions

- Use **black** ink or ball-point pen.
- **Fill in the boxes** at the top of this page with your name, centre number and candidate number.
- Answer **all** questions.
- Answer the questions in the spaces provided
  – *there may be more space than you need.*
- **Calculators may not be used.**
- Diagrams are **NOT** accurately drawn, unless otherwise indicated.
- You must **show all your working out**.

Information

- The total mark for this paper is
- The marks for **each** question are shown in brackets
  – *use this as a guide as to how much time to spend on each question.*

Advice

- Read each question carefully before you start to answer it.
- Keep an eye on the time.
- Try to answer every question.
- Check your answers if you have time at the end.

Turn over ▶

© Peter Bland

1(a) Work out 7500 + 1500

Write your answer in words.

....
(2)

1(b) Write 4748 to the nearest hundred.

....
(1)

1(c) What is the value of the digit 5 in the number 425 986?

....
(1)

1(d) Write down the positive square root of 121.

....
(1)

1(e) Which of these is equal to one million?
Circle your answer.

10^3 10^4 10^5 10^6 10^7

(1)

(Total for Question 1 is 6 marks)

2 Use the numbers from this list to answer the questions.

    5   12   17   25   28   30   42   49

(a) Write down all the multiples of 5.

....

(2)

(b) Write down all the factors of 100.

....

(2)

(c) Write down a square number.

....

(1)

(d) Write down three numbers that have a sum of 60.

.... and.................... and......................

(1)

(Total for Question 2 is 6 marks)

3 Here are two numbers.

            forty thousand   7500

Which number is bigger?

Give a reason for your answer.

Bigger number ....

Reason ....

....

(2)

(Total for Question 3 is 2 marks)

4   *w*, *x* and *y* are three positive whole numbers. *w* is one-fifth of *y*.

x is one-sixth of *y*.

y is less than 100.

What values could *y* take?

....

(5)

(Total for Question 4 is 5 marks)

5   The numbers 13 and 17 are consecutive prime numbers.

The number halfway between them is 15.

15 is **not** a square number.

Find a pair of consecutive prime numbers less than 30 where the number halfway between them is a square number.

.... and ....

(2)

(Total for Question 5 is 2 marks)

6 Work out

$$8^2 \div 4^3$$

....

(2)

(Total for Question 6 is 2 marks)

7 You are given that $34.7 \times 25 = 867.5$

(a) Write down the value of 347×25

....

(1)

(b) Write down the value of $86.75 \div 25$

....

(1)

(c) Work out the value of 34.7×26

....

(2)

(Total for Question 7 is 4 marks)

238

8 A tin of baked beans costs 30p.

A shop has a special offer on the baked beans.

Special offer

Pay for 2 tins and get 1 tin free

30p   30p   Free

Helen wants 12 tins of baked beans.

(a) Work out how much she pays.

£ ....
(3)

The normal price of a toaster is £30

In a sale, the price of the toaster is reduced by 15%.

(b) Work out the sale price of the toaster.

£ ....
(3)

(Total for Question 8 is 6 marks)

9 Work out $\frac{1}{5} + \frac{2}{7}$

....

(Total for Question 9 is 2 marks)

10

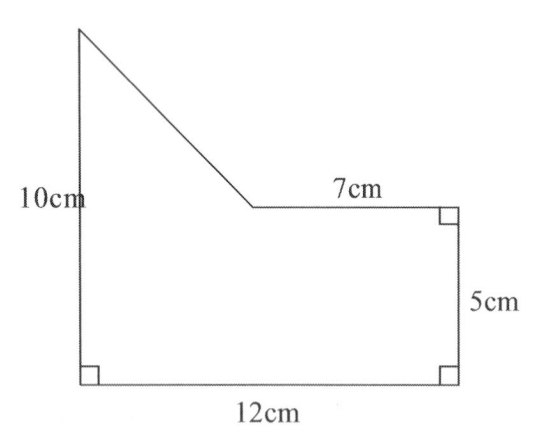

Diagram **NOT** accurately drawn

10cm
7cm
5cm
12cm

Work out the area of the shape.

.... cm²

(Total for Question 10 is 4 marks)

11 Use the information that

$$324 \times 46 = 14904$$

to find the value of

(a) 3.24×4.6

....
(1)

(b) 0.324×0.46

....
(1)

(c) $14904 \div 4.6$

....
(1)

(Total for Question 11 is 3 marks)

12   $2x^2 = 72$

(a) Find a value of x.

....
(2)

(b) Express 72 as a product of its prime factors.

....
(2)

(Total for Question 12 is 4 marks)

13   Here are the ingredients needed to make 8 pancakes.

Pancakes
Ingredients to make **8** pancakes

300 ml milk
1 egg
120 g flour
5 g butter

David makes 24 pancakes.

(a) Work out how much milk he needs.

.... ml
(2)

Louis makes 12 pancakes.

(b) Work out how much flour he needs.

.... g
(2)

(Total for Question 13 is 4 marks)

14 Shagufta has a part-time job.

She is paid £5.60 for each hour she works.

Last week Shagufta worked for 24 hours.

Work out Shagufta's total pay for last week.

£ ....

(Total for Question 14 is 3 marks)

15 Here are the ages, in years, of 15 teachers.

34   53   41   28   37

22   32   40   50   34

44   28   45   45   55

Draw an ordered stem and leaf diagram to show this information.
You must include a key.

Key:

(Total for Question 15 is 3 marks)

16 Using the information that
$$4.8 \times 36 = 172.8$$
write down the value of

(a) 48×36

....
(1)

(b) 4.8×3.6

....
(1)

(c) $172.8 \div 48$

....
(1)

(Total for Question 16 is 3 marks)

17 This rule is used to work out the total cost, in pounds, of hiring a bicycle.

> Multiply the number of days' hire by 3
>
> Add 6 to your answer

Peter hires a bicycle.

The total cost is £18

(a) Work out for how many days he hires the bicycle.

.... days   (2)

(b) Write down an expression, in terms of n, for the total cost, in pounds, of hiring a bicycle for n days.

....
(2)

(Total for Question 17 is 4 marks)

18.

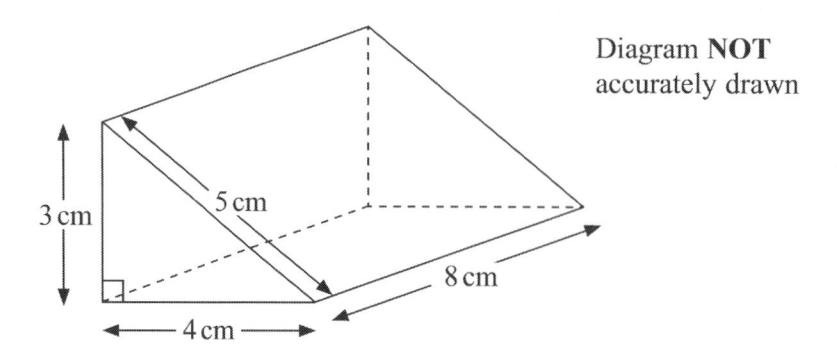

Diagram **NOT** accurately drawn

Work out the total surface area of the triangular prism.
Give the units with your answer.

....

(Total for Question 18 is 4 marks)

19 Work out an estimate for $\dfrac{302 \times 9.96}{0.51}$

(Total for Question 19 is 3 marks)

20 Here is a 4-sided spinner.

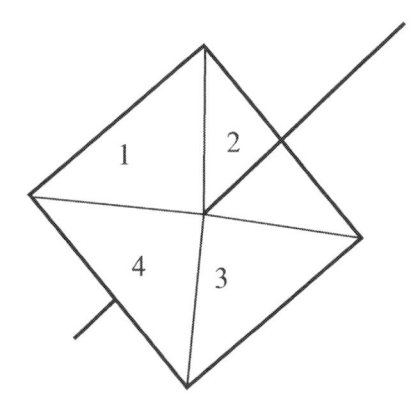

The sides of the spinner are labelled 1, 2, 3 and 4.
The spinner is biased.
The table shows the probability that the spinner will land on each of the colours 1, 4 and 3.

Colour	1	2	3	4
Probability	0.2		0.3	0.1

Work out the probability the spinner will land on 2.

(Total for Question 20 is 2 marks)

21 (a) Write down the reciprocal of 5

....

(b) Work out the value of $2\frac{4}{5} - 1\frac{3}{4}$

Give your answer as a fraction in its simplest form.

....
(3)

(c) Derek says that $4\frac{1}{3}$ is equal to 4.3

Derek is **wrong**.

Explain why.

....

....
(1)

(Total for Question 21 is 5 marks)

22 Write the following numbers in order of size.
Start with the smallest number.

        0.61   0.1   0.16   0.106

....

(Total for Question 22 is 1 mark)

23 Write 0.037 as a fraction.

....

(Total for Question 23 is 1 mark)

24 Write down the 20th odd number.

....

(Total for Question 24 is 1 mark)

25 Write down all the factors of 20

....

(Total for Question 25 is 2 marks)

26 Tara needs to buy chocolate bars for all the children in Year 7
Each of the 130 children get one chocolate bar.

There are 8 chocolate bars in each packet.

Work out the least number of packets of chocolate bars that Tara needs to buy.

....

(Total for Question 26 is 3 marks)

27 Sophie has three tiles.
Each tile has a different number on it.
Sophie puts the three tiles down to make a
number. Each number is made with all three tiles.

How many different numbers can Sophie make?

(Total for Question 27 is 2 marks)

28 One day Shagufta earned £60
She worked for 8 hours.

Work out Shagufta's hourly rate of pay.

£......................

(Total for Question 28 is 2 marks)

29 Work out 15% of 80

....

(Total for Question 29 is 2 marks)

30 There are 3 red beads and 1 blue bead in a jar.
A bead is taken at random from the jar.

What is the probability that the bead is blue?

....

(Total for Question 30 is 1 mark)

31 There are only black pens and green pens in a box.
The ratio of the number of black pens in the box to the number of green pens in the box is 2 : 5

What fraction of the pens are black?

....

(Total for Question 31 is 1 mark)

32 Ron organised an event for a charity.

Each ticket for the event cost £19.95
Ron sold 395 tickets.

Ron paid costs of £6000
He gave all money left to the charity.

(a) Work out an estimate for the amount of money Ron gave to the charity.

£..........................
(3)

(b) Is your answer to (a) an underestimate or an overestimate?
Give a reason for your answer.

....

....

(1)

(Total for Question 32 is 4 marks)

33 Here are some properties of numbers.

      A Even
      B Odd
      C Prime
      D Square
      E Two-digit

(a) Which **two** properties does the number 4 have?

Circle the correct letters.

    A   B   C   D   E

(b) Can one number have **all** of the properties?

Tick a box.

☐ Yes   ☐ No   ☐ Cannot tell

Give a reason for your answer.

....

....

(2)

(c) Write down a number with **three** of the properties.

State which properties it has.

Number ....

Properties .... , ... , ...

(2)

(Total for Question 33 is 4 marks)

34 (a) Round 27 146 correct to

(i) the nearest ten,

.... (1)

(ii) the nearest thousand.

.... (1)

(b) The width of a bench, b, is 984.8 cm correct to one decimal place.

Write down the error interval for the width of the bench.

.... ≤ b < ....
(2)

(c) (i) Write 856 000 000 in standard form.

.... (1)

(ii) Write 4.31×10^{-3} as an ordinary number.

.... (1)

(d) Work out.

$\sqrt[3]{27} + \sqrt{25}$

.... (2)

(Total for Question 34 is 8 marks)

35 (a) Insert one of $<$, $>$ or $=$ to make each statement true.

(i) $^-5$ .... $^-7$ (1)

(ii) 0.09 .... 0.8 (1)

(iii) 6^2 .... 12 (1)

(b) Work out the value of $5^2 \times 10^2$.

(b) ....

(2)

(Total for Question 35 is 5 mark)

36 (a) Write numbers in the boxes below to make the statement true.

$$15 \times 20 = 5 \times \boxed{} = 6 \times \boxed{}$$

(2)

(b) Louis thinks of a number.
If he cubes his number and then adds 9, he gets 17.

What number is he thinking of?

(b) ....

(2)

(Total for Question 36 is 4 mark)

Write your name here

Surname

Other names

In the style of:
Pearson Edexcel
Level 1/Level 2 GCSE (9 - 1)

Centre Number

Candidate Number

Mathematics

Probability

Higher Tier

GCSE style questions arranged by topic

Paper Reference
1MA1/2H

You must have: Ruler graduated in centimetres and millimetres, protractor, pair of compasses, pen, HB pencil, eraser, calculator.

Total Marks

Instructions

- Use **black** ink or ball-point pen.
- **Fill in the boxes** at the top of this page with your name, centre number and candidate number.
- Answer **all** questions.
- Answer the questions in the spaces provided
  – *there may be more space than you need.*
- **Calculators may be used.**
- If your calculator does not have a π button, take the value of π to be 3.142 unless the question instructs otherwise.
- Diagrams are **NOT** accurately drawn, unless otherwise indicated.
- You must **show all your working out**.

Information

- The total mark for this paper is
- The marks for **each** question are shown in brackets
  – *use this as a guide as to how much time to spend on each question.*

Advice

- Read each question carefully before you start to answer it.
- Keep an eye on the time.
- Try to answer every question.
- Check your answers if you have time at the end.

Turn over

© Peter Bland

1 David goes to a club.
   He has one go at Darts.
   He has one go at Pool.

   The probability that he wins at Darts is 0.3
   The probability that he wins at Pool is 0.4

   (a) Complete the probability tree diagram.

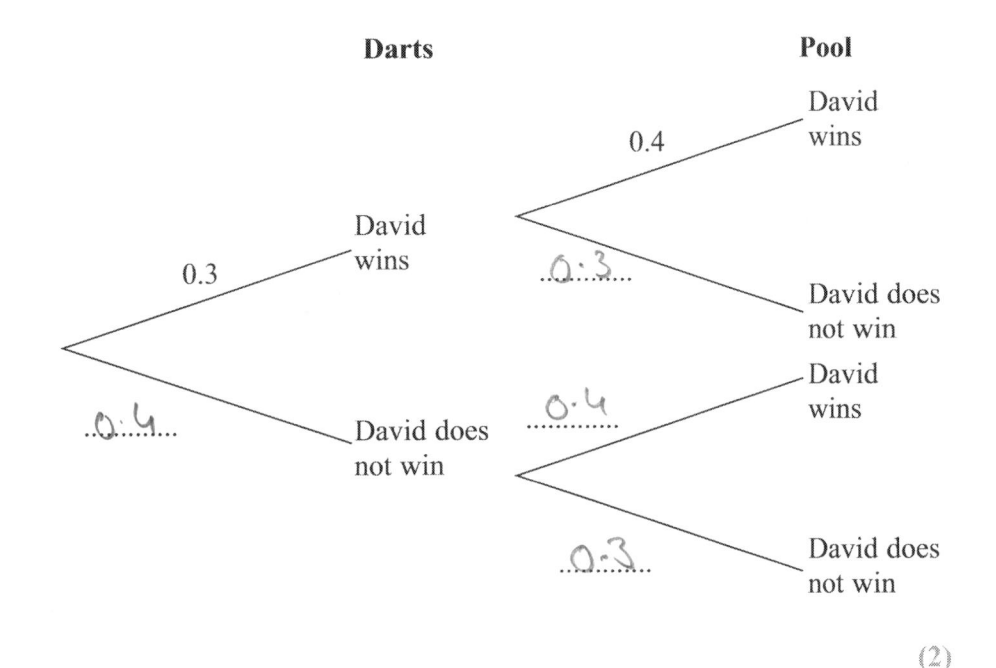

                    Darts   **Pool**

                                            0.4   David wins
                         David wins
                                        ...0.3...   David does not win
            0.3
                                            0.4   David wins
           ...0.4... David does not win
                                        ...0.3...   David does not win

                                                                (2)

   (b) Work out the probability that David wins at Darts and also wins at Pool.

   0.3×0.4
   $= 0.12$

                                                                (2)

                                   Total for Question 1 is 4 marks)

2 A bowl contains 3 oranges, 5 mangoes and 7 bananas.

One fruit is taken, at random, from the bowl and **not** replaced. Another fruit is then taken, at random, from the bowl.
A tree diagram representing these two events is shown below.

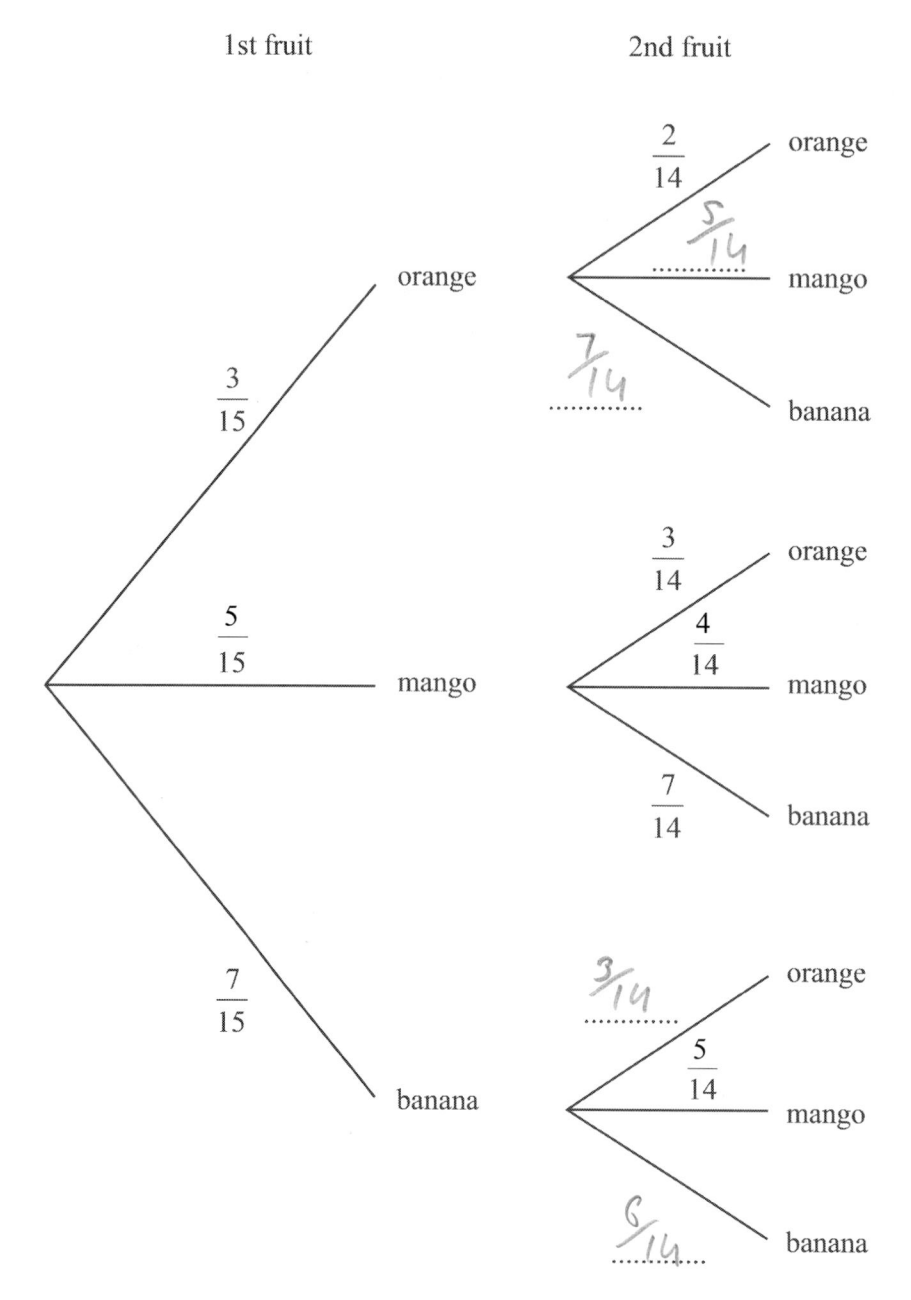

1st fruit   2nd fruit

- $\frac{3}{15}$ orange
  - $\frac{2}{14}$ orange
  - $\frac{5}{14}$ mango
  - $\frac{7}{14}$ banana
- $\frac{5}{15}$ mango
  - $\frac{3}{14}$ orange
  - $\frac{4}{14}$ mango
  - $\frac{7}{14}$ banana
- $\frac{7}{15}$ banana
  - $\frac{3}{14}$ orange
  - $\frac{5}{14}$ mango
  - $\frac{6}{14}$ banana

(a) Complete the tree diagram representing these two events.

(2)

(b) Find the probability that both fruit are bananas. Give your answer as a simplified fraction.

$\frac{7}{15} \times \frac{6}{14} = \frac{42}{210} \quad \frac{21}{105} = \frac{3}{15}$

$\frac{3}{15}$

(2)

(Total for Question 2 is 4 marks)

3 Tara has 8 balls in a box.

5 of the balls are blue.
3 of the balls are red.

Tara takes at random a ball from the box and writes down its colour.
Tara puts the ball back in the box.

Then Tara takes at random a second ball from the box, and writes down its colour.

(a) Complete the probability tree diagram.

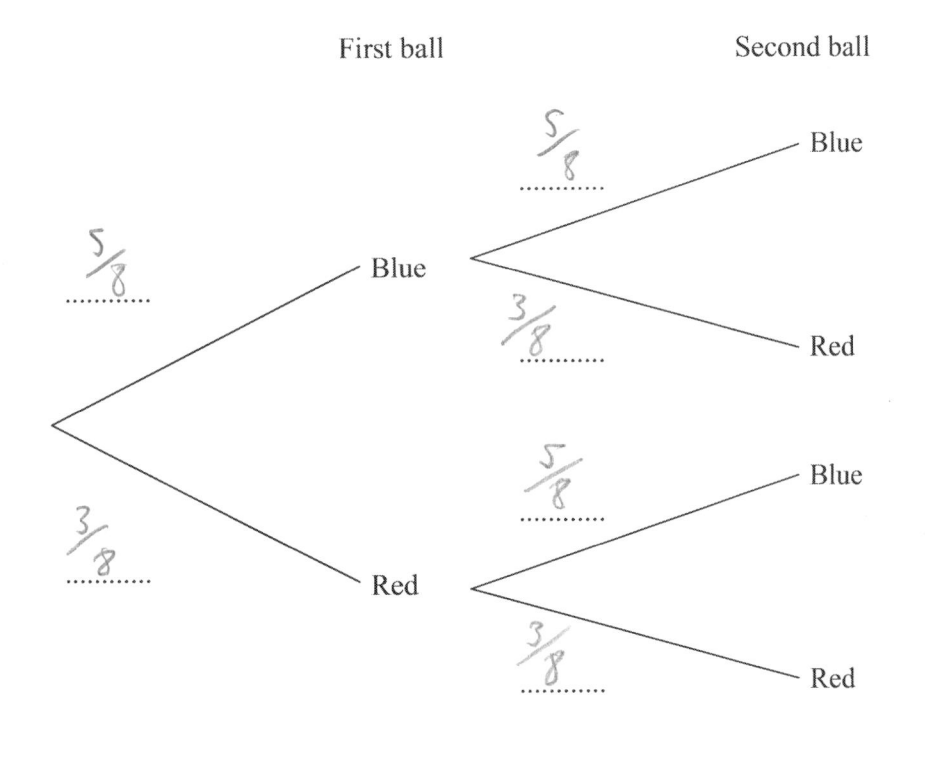

First ball ..... Second ball

- First ball Blue: $\frac{5}{8}$
- First ball Red: $\frac{3}{8}$
- Blue → Blue: $\frac{5}{8}$
- Blue → Red: $\frac{3}{8}$
- Red → Blue: $\frac{5}{8}$
- Red → Red: $\frac{3}{8}$

(2)

(b) Work out the probability that Tara takes exactly one ball of each colour from the box.

$$\frac{5}{8} \times \frac{3}{8} = \frac{15}{64}$$

$\frac{15}{64}$

(3)

(Total for Question 3 is 5 marks)

4. In a game of chess, a player can either win, draw or lose.

The probability that Sophie wins any game of chess is 0.5

The probability that Sophie draws any game of chess is 0.2

Sophie plays 2 games of chess.

(a) Complete the probability tree diagram.

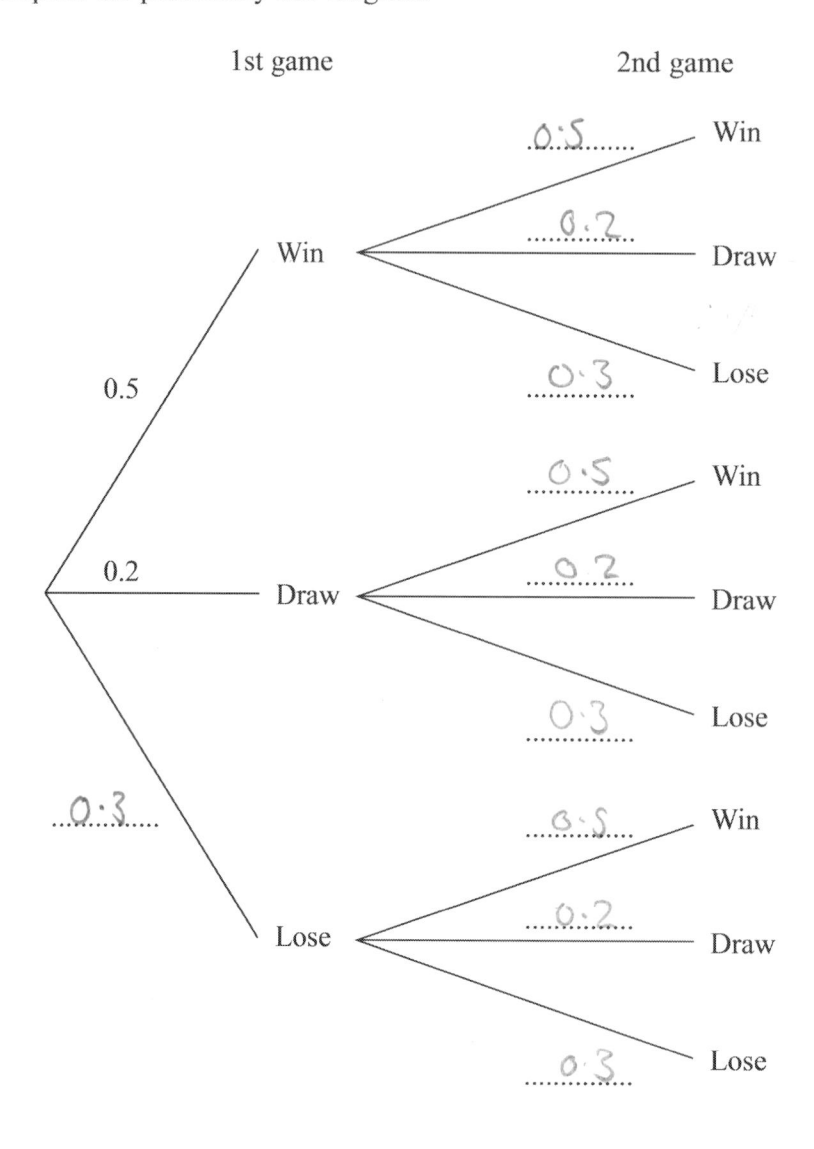

1st game	2nd game

- 0.5 → Win
  - 0.5 → Win
  - 0.2 → Draw
  - 0.3 → Lose
- 0.2 → Draw
  - 0.5 → Win
  - 0.2 → Draw
  - 0.3 → Lose
- 0.3 → Lose
  - 0.5 → Win
  - 0.2 → Draw
  - 0.3 → Lose

(2)

(b) Work out the probability that Sophie will win both games.

0.5 × 0.5 = 0.25

.......0.25.......

(2)

(Total for Question 4 is 4 marks)

5 Louis puts 3 red balls and 4 blue balls in a bag.
He takes at random a ball from the bag.
He writes down the colour of the ball.
He puts the ball in the bag again.
He then takes at random a second ball from the bag.

(a) Complete the probability tree diagram.

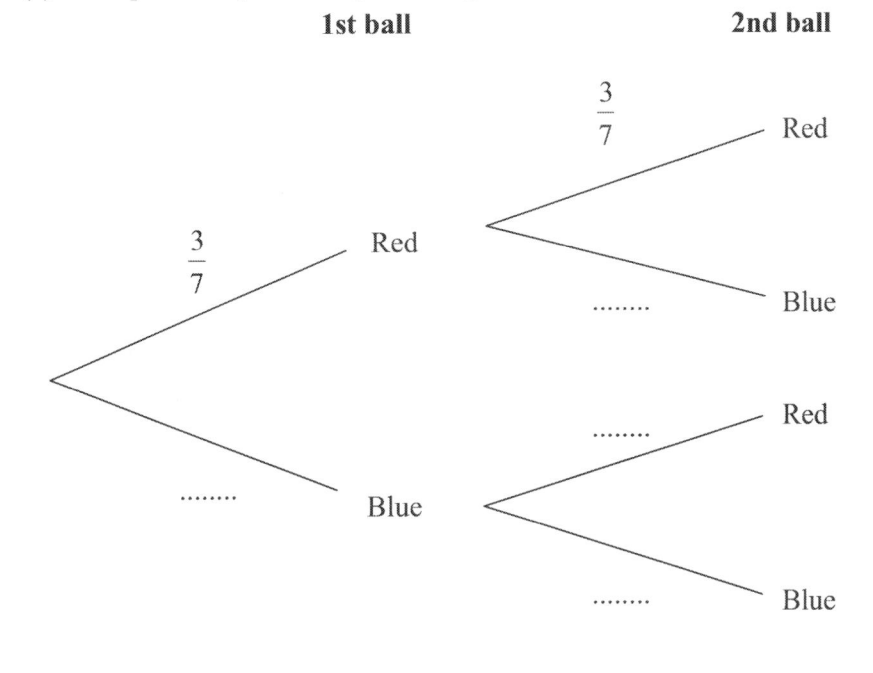

1st ball   **2nd ball**

$\frac{3}{7}$ Red

$\frac{3}{7}$ Red

.... Blue

.... Red

Blue

.... Blue

(2)

(b) Work out the probability that Louis takes two red balls.

....
(2)

(Total for Question 5 is 4 marks)

6 Helen and Anthony each take a medical.

The probability that Helen will pass the medical is 0.9 The probability that Anthony will pass the medical is 0.7

(a) Complete the probability tree diagram.

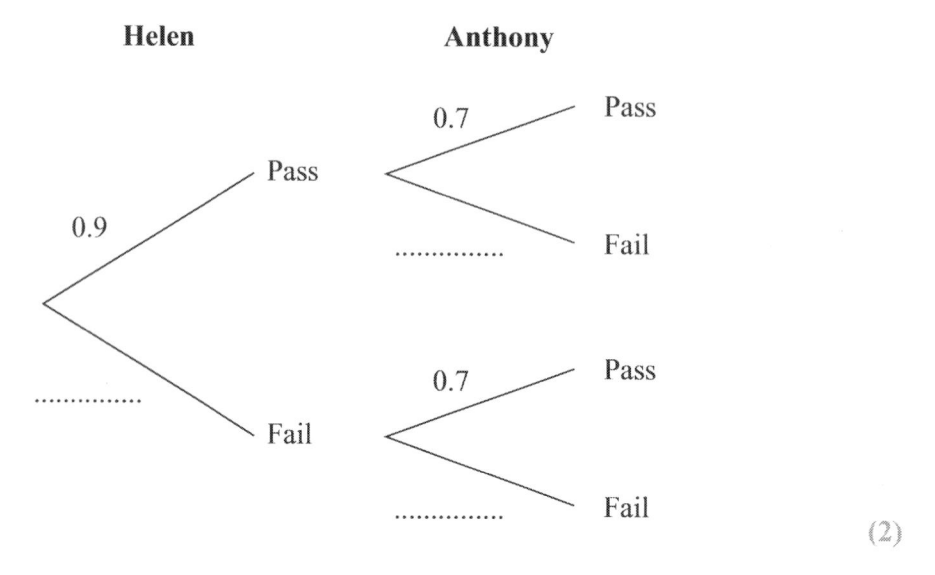

        Helen   **Anthony**

```
                    0.7 ─── Pass
            Pass ──<
           /         ......── Fail
      0.9/
        /
        \
      ...\
          \         0.7 ─── Pass
            Fail ──<
                    ......── Fail
```

(2)

(b) Work out the probability that both Helen and Anthony will pass the medical.

....

(2)

(c) Work out the probability that only one of them will pass the medical.

....

(3)

(Total for Question 6 is 7 marks)

7 There are 3 red sweets, 2 purple sweets and 5 orange sweets in a bag.

Georgina takes a sweet at random.
She eats the sweet.
She then takes another sweet at random.

Work out the probability that both the sweets are the same colour.

....

(Total for Question 7 is 4 marks)

8 Sheila is going to play one game of darts and one game of dominoes.

The probability that she will win the game of darts is $\frac{3}{4}$

The probability that she will win the game of dominoes $\frac{1}{3}$

is (a) Complete the probability tree diagram.

darts **dominoes**

```
         ...3/4...  Sheila wins
                        ...1/3...  Sheila wins
                        .........  Sheila does not win
         .........  Sheila does not win
                        .........  Sheila wins
                        .........  Sheila does not win
```

(2)

(b) Work out the probability that Sheila will win **exactly** one game.

....
(3)

Sheila played one game of darts and one game of dominoes on a number of Fridays.
She won at **both** darts and dominoes on 21 Fridays.

(c) Work out an estimate for the number of Fridays on which Sheila did not win either game.

....
(3)

(Total for Question 8 is 8 marks)

9 Ron plays one game of golf and one game of pool.

The probability that Ron will win at golf is $\frac{3}{4}$

The probability that Ron will win at pool is $\frac{1}{3}$

(a) Complete the probability tree diagram below.

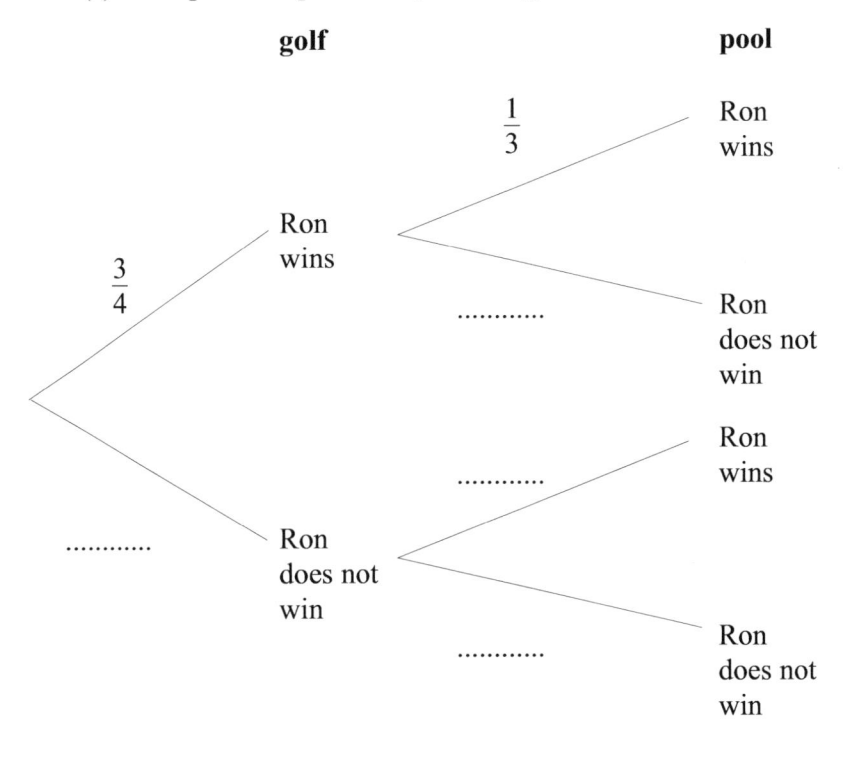

golf   pool

- $\frac{3}{4}$ Ron wins
  - $\frac{1}{3}$ Ron wins
  - .... Ron does not win
- .... Ron does not win
  - .... Ron wins
  - .... Ron does not win

(2)

(b) Work out the probability that Ron wins both games.

....
(2)

(c) Work out the probability that Ron will win only one game.

....
(3)

(Total for Question 9 is 7 marks)

10 A and B are two sets of traffic lights on a road.

The probability that a car is stopped by lights A is 0.4

If a car is stopped by lights A, then the probability that the car is **not** stopped by lights B is 0.7

If a car is **not** stopped by lights A, then the probability that the car is **not** stopped by lights B is 0.2

(a) Complete the probability tree diagram for this information.

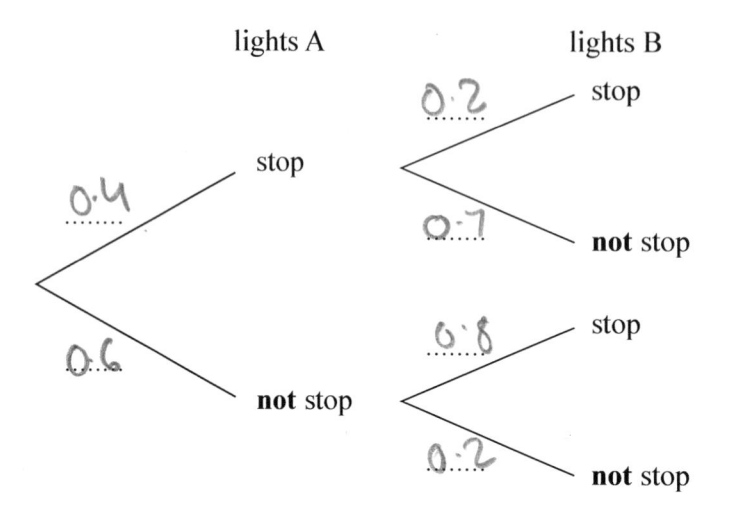

```
         lights A              lights B
                          0.2  ─── stop
              stop   <
      0.4             0.7  ─── not stop
   <
      0.6             0.8  ─── stop
            not stop <
                          0.2  ─── not stop
```

(2)

Derek drove along this road.
He was stopped by just one of the sets of traffic lights.

(b) Is it more likely that he was stopped by lights A or by lights B?
You must show your working.

lights A = 0.4 × 0.2 = 0.8

lights B = 0.6 × 0.2 = 0.12

lights B is more likely.

(3)

(Total for Question 10 is 5 marks)

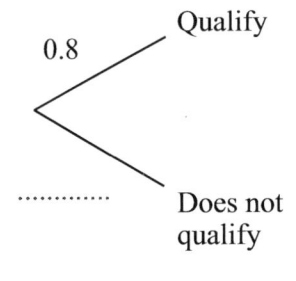

11 On Friday, Arshan takes part in a long jump competition.

He has to jump at least 7.5 metres to qualify for the final on Saturday.

- He has up to three jumps to qualify.
- If he jumps at least 7.5 metres he does **not** jump again on Friday.

Each time Arshan jumps, the probability he jumps at least 7.5 metres is 0.8 Assume each jump is independent.

(a) Complete the tree diagram.

First jump   **Second jump**   **Third jump**

0.8 — Qualify

.... — Does not qualify

(2)

(b) Work out the probability that he does **not** need the third jump to qualify.

(2)

(Total for Question 11 is 4 marks)

12 Nikki goes to a fun fair.
She has one go at Hoopla.
She has one go on the Coconut shy.

The probability that she wins at Hoopla is 0.4
The probability that she wins on the Coconut shy is 0.3

(a) Complete the probability tree diagram.

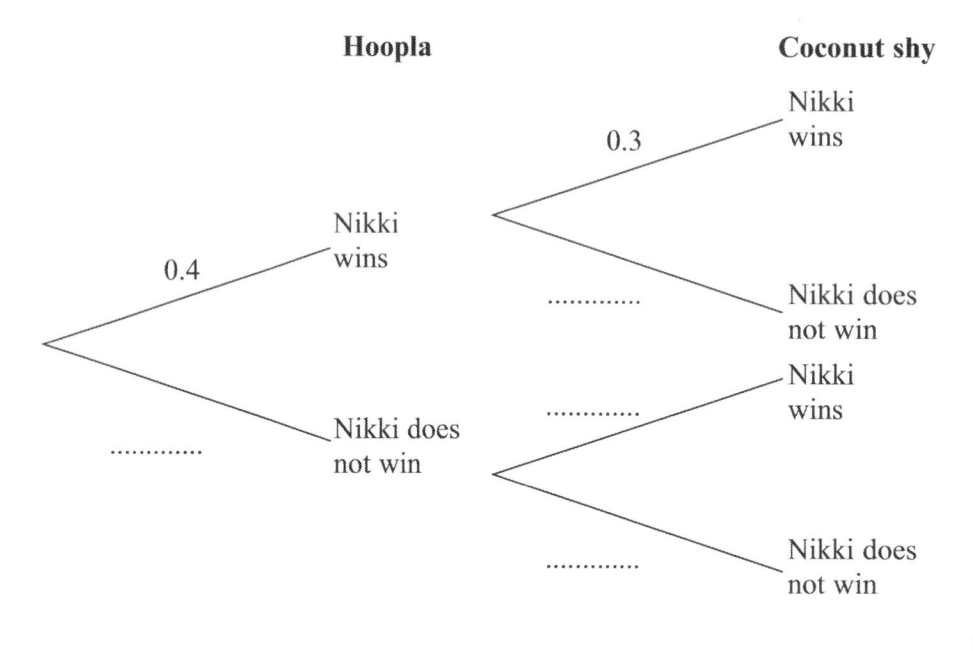

Hoopla   **Coconut shy**

0.4 — Nikki wins
    0.3 — Nikki wins
    .... — Nikki does not win

.... — Nikki does not win
    .... — Nikki wins
    .... — Nikki does not win

(2)

(b) Work out the probability that Nikki wins at Hoopla and also wins on the Coconut shy.

....

(2)

(Total for Question 12 is 4 marks)

13 Noah has 10 coins in a bag.
There are three £1 coins and seven 50 pence coins.

Noah takes at random, 3 coins from the bag.

Work out the probability that he takes exactly £2.50

....

(Total for Question 13 is 4 marks)

14 Some of the children at a nursery arrive by car.

- 40% of the children at the nursery are boys.
- 70% of the boys at the nursery arrive by car.
- 60% of the girls at the nursery arrive by car.

What is the probability that a child chosen at random from the nursery arrives by car?

....
(5)

(Total for Question 14 is 5 marks)

15 Peter believes he knows what his brother John is thinking.
He carries out an experiment to test this.

Peter and John sit back-to-back.
John rolls an ordinary fair dice.
John then thinks about the number on the dice while Peter tries to predict this number.

(a) In 300 attempts, how many correct predictions would you expect Peter to make if he was just guessing?

(a) ....
(2)

(b) The results of the first 15 attempts are shown in the table.

John's number	2	6	5	3	2	1	5	1	3	4	4	6	1	6	5
Peter's prediction	2	4	3	1	2	6	1	6	4	3	2	6	5	2	3
Matching pair	✓				✓							✓			

Estimate the probability of getting a matching pair using the results of

(i) the first five attempts,

(b)(i) ....
(1)

(ii) all 15 attempts.

(ii) ....
(1)

(c) Use answers from **(a)** and **(b)** to comment on Peter's belief that he knows what John is thinking.

(2)

(Total for Question 15 is 6 marks)

16 A coin is rolled onto a grid of squares.

It lands randomly on the grid.

To win, the coin must land completely within one of the squares.

Misbah and Hamera each roll the coin a number of times and record their results.

	Number of wins	Number of losses
Misbah	6	44
Hamera	28	72

(a) Work out **two** different estimates for the probability of winning.

.... and ....

(2)

(b) Which of your estimates is the better estimate for the probability of winning?
Give a reason for your answer.

Answer ....

Reason ....

....

....

(1)

(Total for Question 16 is 3 marks)

17 Bag X contains 9 blue balls and 18 red balls.

Bag Y contains 7 blue balls and 14 red balls.

Helen picks a ball at random from bag X. She puts the ball into bag Y.

Matt now picks a ball at random from bag Y.

Show that:

   P (Helen picks a blue ball) = P (Matt picks a blue ball)

(4)

(Total for Question 17 is 4 marks)

Write your name here

Surname

Other names

In the style of:
Pearson Edexcel
Level 1/Level 2 GCSE (9 - 1)

Centre Number

Candidate Number

Mathematics

Quadratics

Higher Tier

GCSE style questions arranged by topic

Paper Reference
1MA1/2H

You must have: Ruler graduated in centimetres and millimetres, protractor, pair of compasses, pen, HB pencil, eraser, calculator.

Total Marks

Instructions

- Use **black** ink or ball-point pen.
- **Fill in the boxes** at the top of this page with your name, centre number and candidate number.
- Answer **all** questions.
- Answer the questions in the spaces provided
  - *there may be more space than you need.*
- **Calculators may be used.**
- If your calculator does not have a π button, take the value of π to be 3.142 unless the question instructs otherwise.
- Diagrams are **NOT** accurately drawn, unless otherwise indicated.
- You must **show all your working out**.

Information

- The total mark for this paper is
- The marks for **each** question are shown in brackets
  - *use this as a guide as to how much time to spend on each question.*

Advice

- Read each question carefully before you start to answer it.
- Keep an eye on the time.
- Try to answer every question.
- Check your answers if you have time at the end.

Turn over

© Peter Bland

1   Simplify fully

$$\frac{6x^2 + x - 1}{4x^2 - 1}$$

....

(Total for Question 1 is 4 marks)

2   Simplify fully $\dfrac{x^2 - 8x + 15}{2x^2 - 7x - 15}$

....

(Total for Question 2 is 3 marks)

3 The diagram below shows a 6-sided shape.
All the corners are right angles.
All the measurements are given in centimetres.

Diagram **NOT** accurately drawn

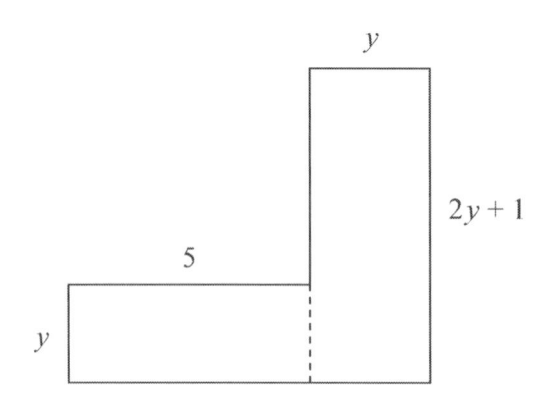

The area of the shape is 95 cm².

(a) Show that $2y^2 + 6y - 95 = 0$

(3)

(b) Solve the equation

$$2y^2 + 6y - 95 = 0$$

Give your solutions correct to 3 significant figures.

(3)

(Total for Question 3 is 6 marks)

$y = $ .... or $y = $ ....

4 (a) Rearrange this equation

$$\frac{5}{x+2} = \frac{4-3x}{x-1}$$

to give $3x^2 + 7x - 13 = 0$

(3)

(b) Solve $3x^2 + 7x - 13 = 0$
correct to 2 decimal places.

$x = $ .... or $x = $ ....

(3)

(Total for Question 4 is 6 marks)

5 (a) Expand and simplify $(x + 3)(x - 2)$

....
(2)

(b) Factorise $x^2 + 7x + 10$

....
(2)

(c) $x = 3y + 4(z - y)$

Find the value of x when $y = 6$ and $z = 5$

$x =$ ....
(3)

(Total for Question 5 is 7 marks)

6 (a) Factorise $x^2 - 7x + 10$

....
(2)

(b) Solve $x^2 - 7x + 10 = 0$

$x = $ ....

or $x = $ ....
(1)

(Total for Question 6 is 3 marks)

7 (a) Simplify $4a + 3c - 2a + c$

....
(1)

(b) $S = \frac{1}{2}at^2$

Find the value of S when $t = 3$ and $a = \frac{1}{4}$

$S = $ ....
(2)

(c) Factorise $x^2 - 5x$

....
(2)

(d) Expand and simplify $(x + 3)(x + 4)$

....
(2)

(e) Factorise $y^2 + 8y + 15$

....
(2)

(Total for Question 7 is 9 marks)

8 (a) Simplify $(c^2 k^5)^4$

....
(1)

(b) Expand and simplify $(3x + 5)(4x - 1)$

....
(2)

(c) Solve $x^2 - 3x - 10 = 0$

$x = $ ....
(3)

(Total for Question 8 is 6 marks)

9 The plan below shows a large rectangle of length $(2x + 6)$ m and width x m.

A smaller rectangle of length x m and width 3 m is cut out and removed.

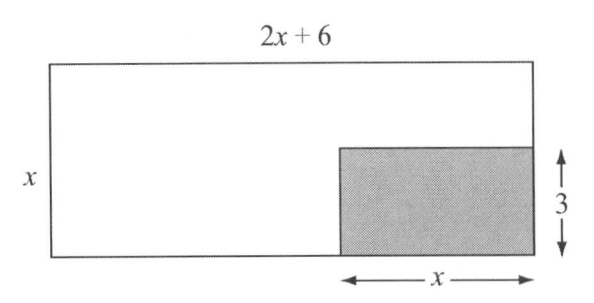

Diagram **NOT** accurately drawn

The area of the shape that is left is 100 m².

(a) Show that $\quad 2x^2 + 3x - 100 = 0$

(3)

(b) Calculate the length of the smaller rectangle.
Give your answer correct to 3 significant figures.

.... m

(4)

(Total for Question 9 is 7 marks)

10 (a) Complete the table of values for $y = x^2 + x - 2$

x	−4	−3	−2	−1	0	1	2
y	10		0	−2			4

(2)

(b) On the grid below, draw the graph of $y = x^2 + x - 2$ for values of x from −4 to 2

(2)

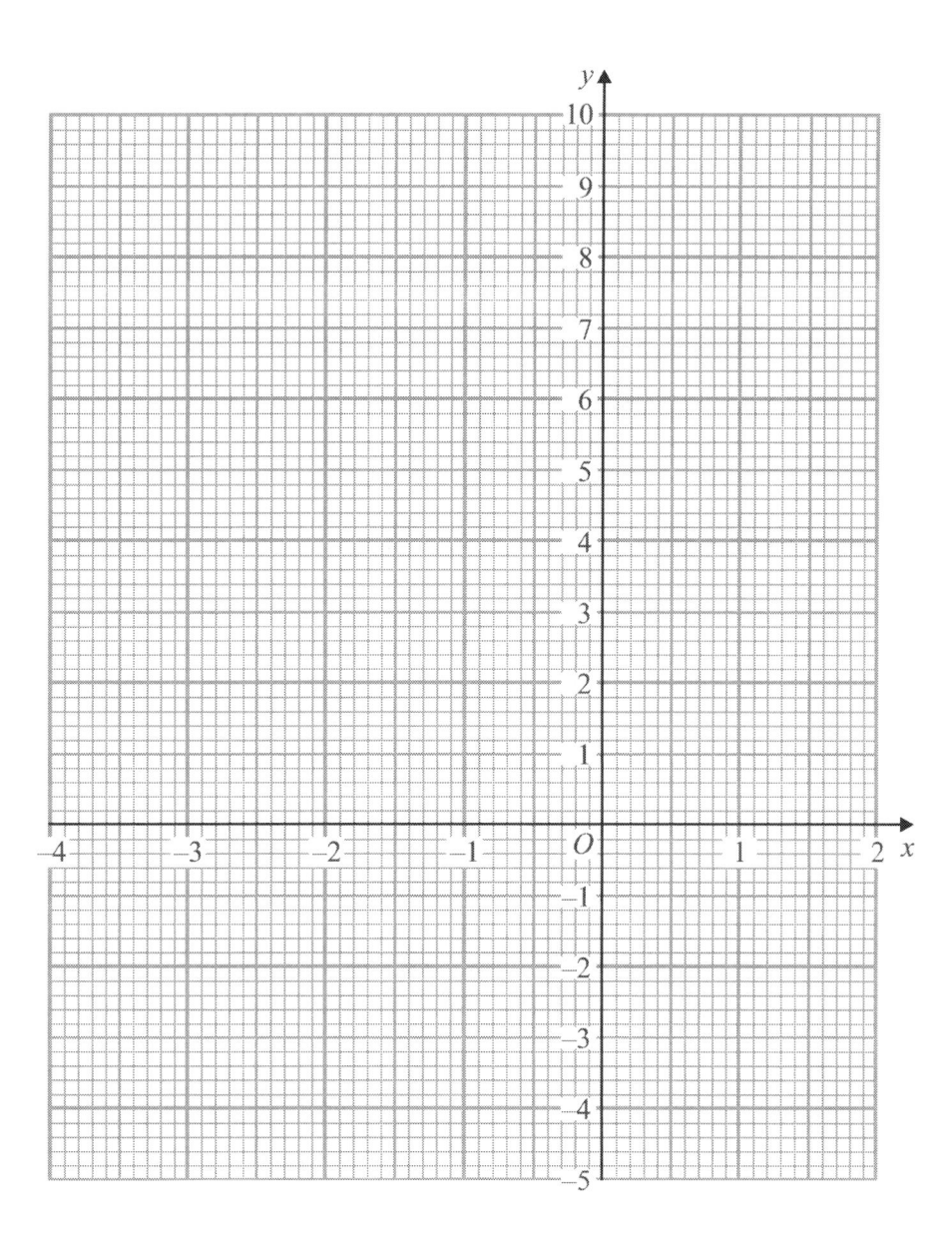

(c) Use your graph to find estimates for the solutions of $x^2 + x - 2 = 0$

$x = $ ....

$x = $ ....

(1)

(Total for Question 10 is 5 marks)

11 (a) Construct the graph of $x^2 + y^2 = 9$

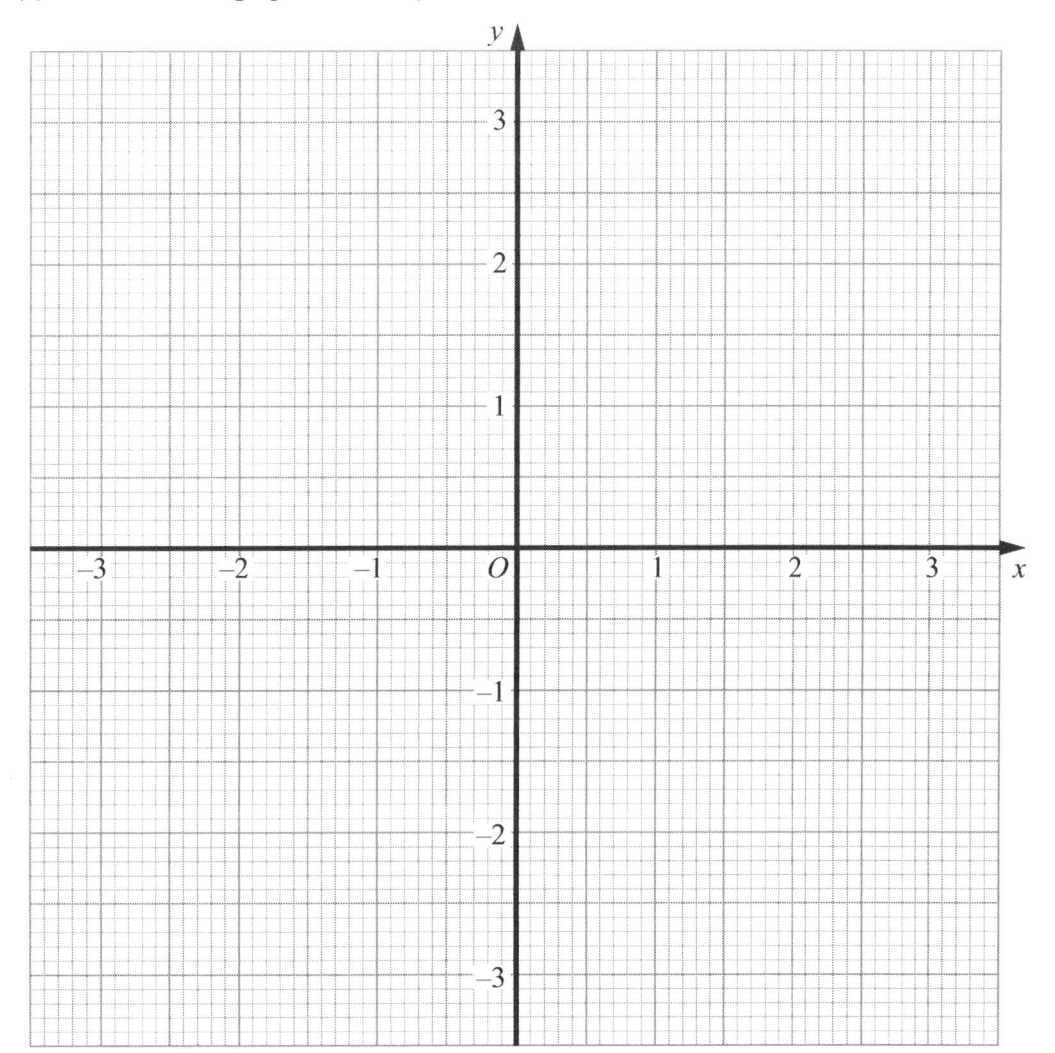

(2)

(b) By drawing the line $x + y = 2$ on the grid, solve the equations $\quad x^2 + y^2 = 9$
$\qquad\qquad\qquad\qquad\qquad\qquad\qquad\qquad\qquad\qquad\qquad\qquad\qquad\qquad x + y = 2$

$x = $........................, $y = $........................

or $x = $........................, $y = $........................

(3)

(Total for Question 11 is 5 marks)

12 (a) Complete the table of values for $y = x^2 - 4x + 1$

x	−1	0	1				
y		1	−2		−2		6

(2)

(b) On the grid, draw the graph of $y = x^2 - 4x + 1$

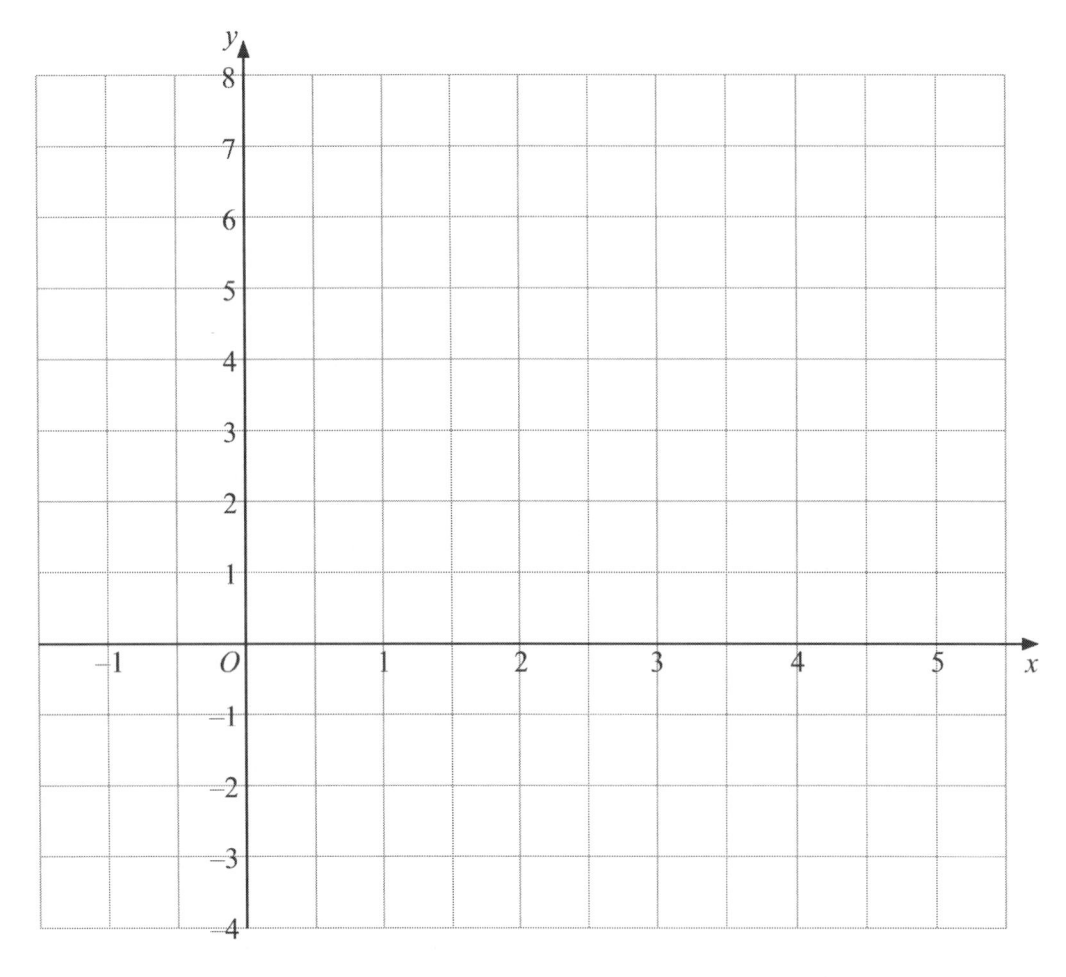

(2)

(Total for Question 12 is 4 marks)

13 (a) Complete the table of values for $y = x^2 - 3x - 1$

x	−2	−1	0				
y		3	1	−3		−1	

(2)

(b) On the grid, draw the graph of $y = x^2 - 3x - 1$ for values of x from −2 to 4

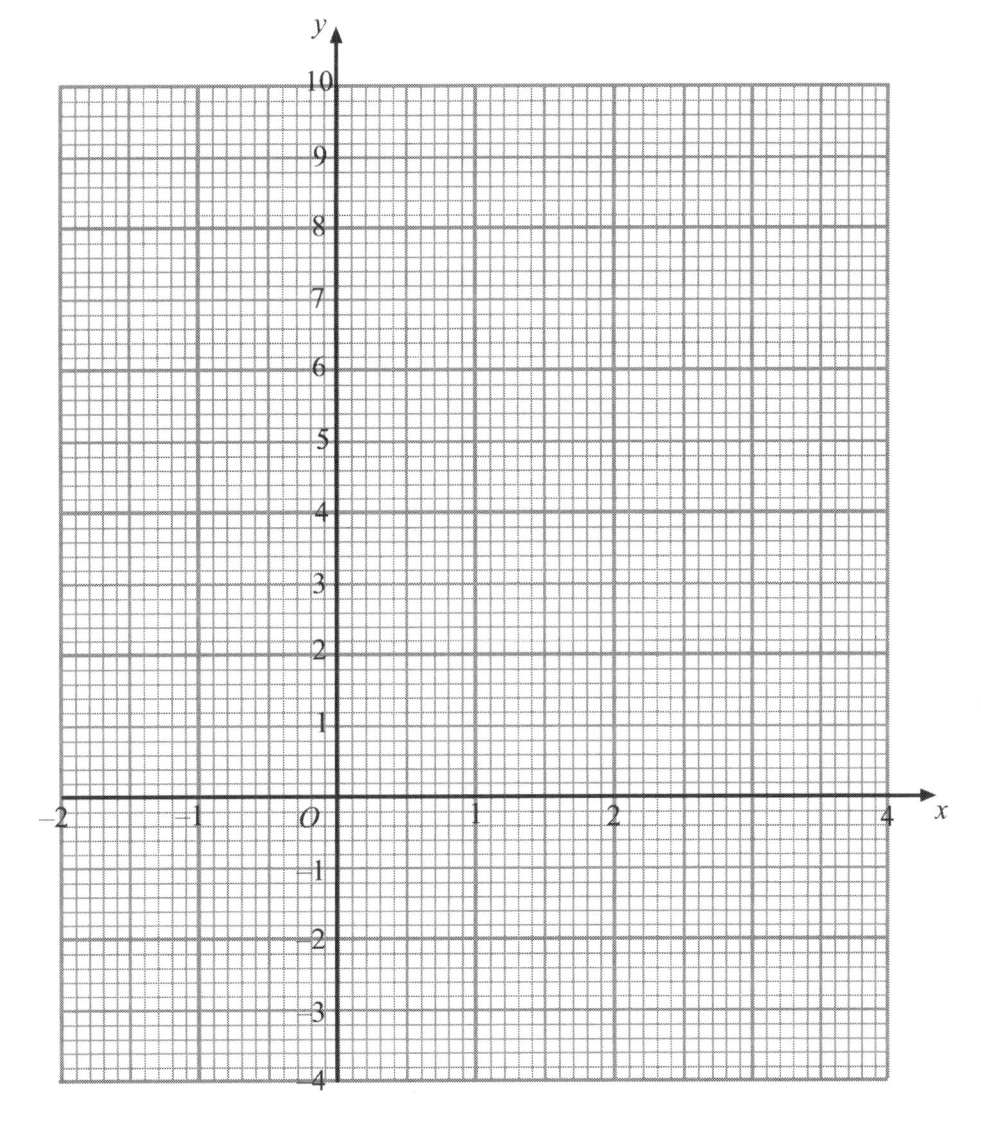

(2)

(Total for Question 13 is 4 marks)

14 Show that any straight line that passes through the point (1, 1) must intersect the curve with equation $x^2 + y^2 = 16$ at two points.

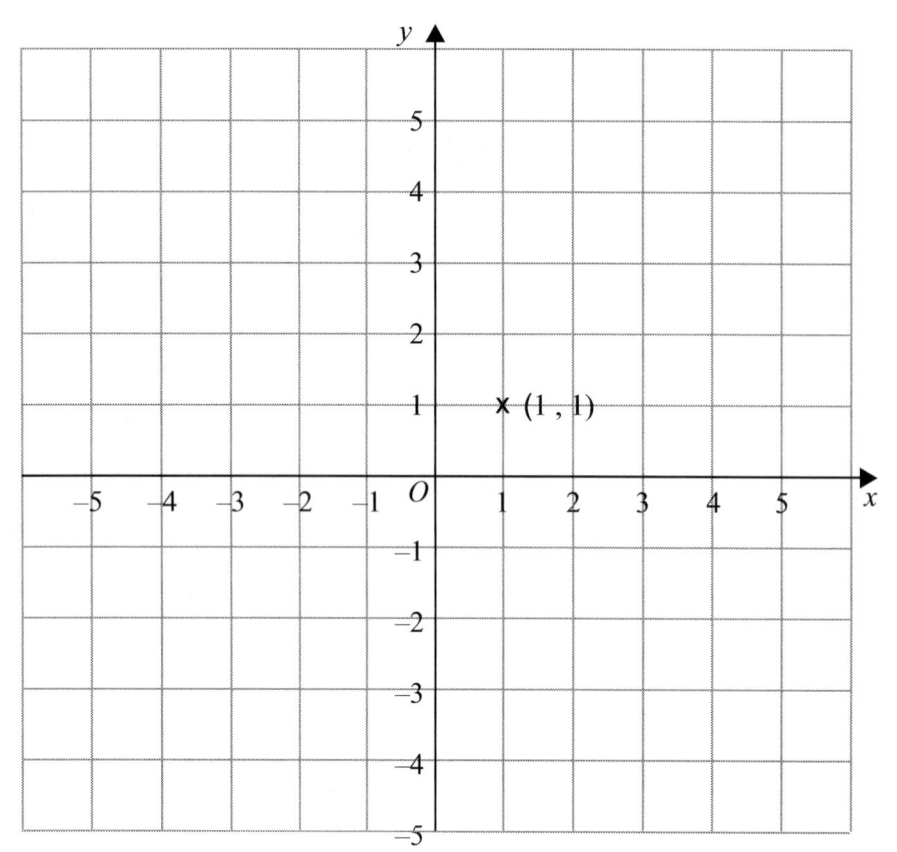

(Total for Question 14 is 3 marks)

15 The graph of y = f(x) is drawn on the grid.

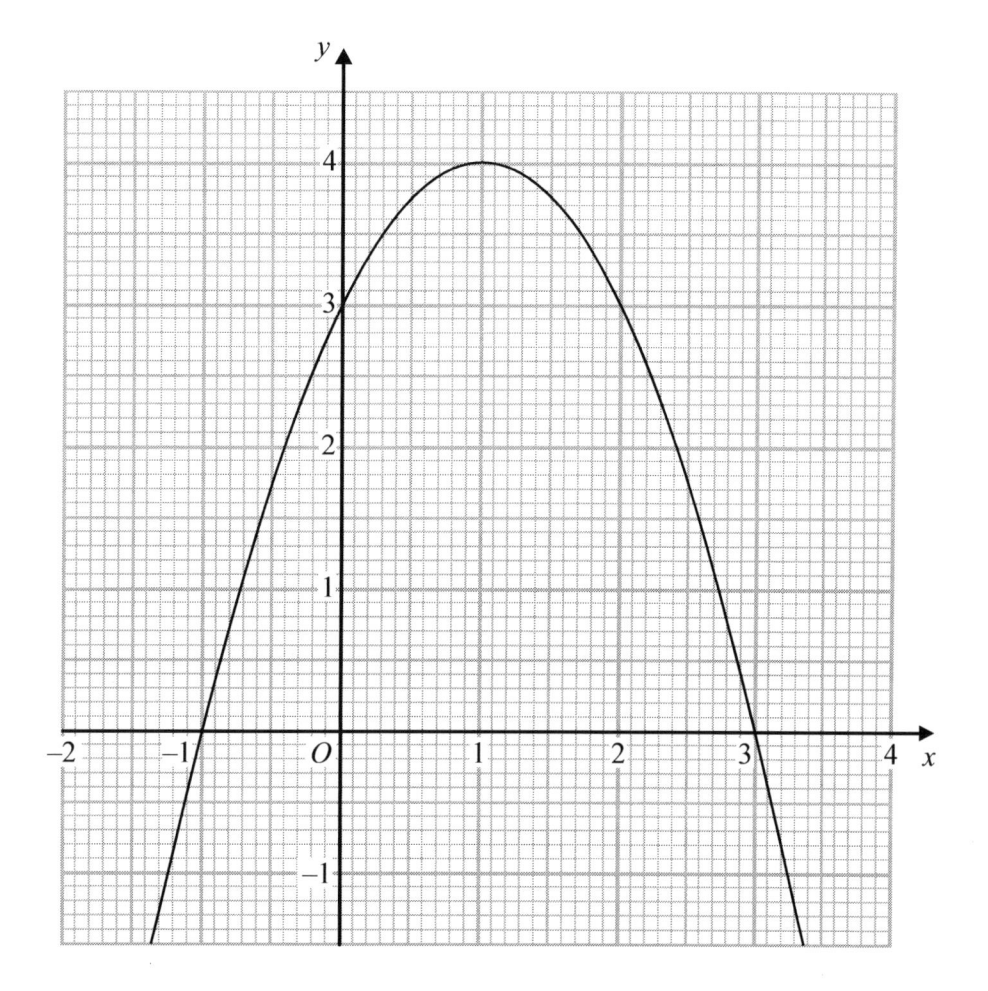

(a) Write down the coordinates of the turning point of the graph.

(1)

(b) Write down the roots of f(x) = 2

(1)

(c) Write down the value of f(0.5)

(1)

(Total for Question 15 is 3 marks)

16 $2x^2 - 6x + 5$ can be written in the form $a(x-b)^2 + c$

where a, b and c are positive numbers.

(a) Work out the values of a, b and c.

$a = $ ....

$b = $ ....

$c = $ ....

(2)

(b) Using your answer to part (a), or otherwise, solve $2x^2 - 6x + 5 = 8.5$

....
(3)

(Total for Question 16 is 5 marks)

17 (a) Write $x^2 + 10x + 29$ in the form $(x+a)^2 + b$

p *q* — always this form.

$(x + \frac{p}{2})^2 - (\frac{x}{2})^2 + q$ ← always $-b$

$(x + \frac{10}{2})^2 - (\frac{10}{2})^2 + 29$

$(x+5)^2 - (25 + 29)$

$(x+5)^2 + 4$

.... $(x+5)^2 + 4$ ....

(3)

(b) Write down the coordinates of the turning point of the graph of $y = x^2 + 10x + 29$.

(.............. , ...)

(1)

(Total for Question 17 is 4 marks)

18 Solve these simultaneous equations algebraically.

$$y = x - 3$$
$$y = 2x^2 + 8x - 7$$

$x = $ .... , $y = $ ....

$x = $ .... , $y = $ ....

(6)

(Total for Question 18 is 6 marks)

Write your name here

Surname

Other names

Pearson Edexcel
Level 1/Level 2 GCSE (9 - 1)

Centre Number

Candidate Number

Mathematics

Scatter Graphs

Higher Tier

GCSE style questions arranged by topic

Paper Reference
1MA1/3H

You must have: Ruler graduated in centimetres and millimetres, protractor, pair of compasses, pen, HB pencil, eraser, calculator.

Total Marks

Instructions

- Use **black** ink or ball-point pen.
- **Fill in the boxes** at the top of this page with your name, centre number and candidate number.
- Answer **all** questions.
- Answer the questions in the spaces provided
  – *there may be more space than you need.*
- **Calculators may be used.**
- If your calculator does not have a π button, take the value of π to be 3.142 unless the question instructs otherwise.
- Diagrams are **NOT** accurately drawn, unless otherwise indicated.
- You must **show all your working out**.

Information

- The total mark for this paper is
- The marks for **each** question are shown in brackets
  – *use this as a guide as to how much time to spend on each question.*

Advice

- Read each question carefully before you start to answer it.
- Keep an eye on the time.
- Try to answer every question.
- Check your answers if you have time at the end.

Turn over

© Peter Bland

1 (a) Andy, Lauren and Noah are playing with a normal fair dice. They each predict the next seven throws.

Andy	1	2	1	2	1	2	1
Lauren	3	5	2	2	4	6	1
Noah	4	4	4	4	4	4	4

Which, if any, of these predictions is the most likely?
Circle your choice and explain your answer.

Andy   Lauren   Noah   All are equally likely

(2)

(b) Nikki makes a six-sided dice.
To test the dice she throws it 100 times.
After each 10 throws she records the number of sixes thrown.
The relative frequencies for the first 90 throws are shown on the graph.

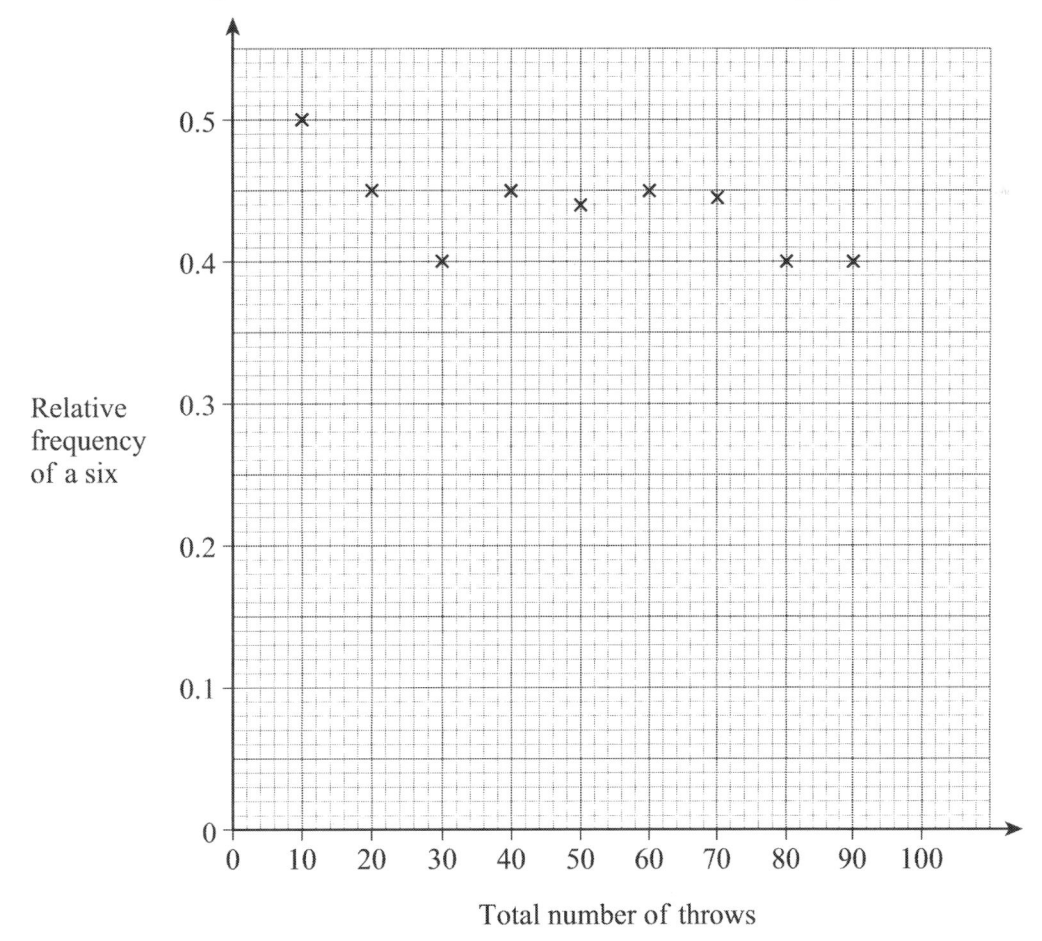

(b) (i) How many sixes were there in the first 10 throws?

(1)

(ii) After 100 throws there were 42 sixes.

Calculate and plot the relative frequency of a six after 100 throws.

(1)

(iii) How many sixes would you expect to get after 100 throws of a **fair** dice?

....

(1)

(iv) Is Nikki's dice fair?
Tick the correct box.

☐ Yes   ☐ No

Give a reason for your answer.

(1)

(Total for Question 1 is 6 marks)

2 The scatter graph shows some information about 10 cars.
It shows the time, in seconds, it takes each car to go from 0 mph to 60 mph.
For each car, it also shows the maximum speed, in mph.

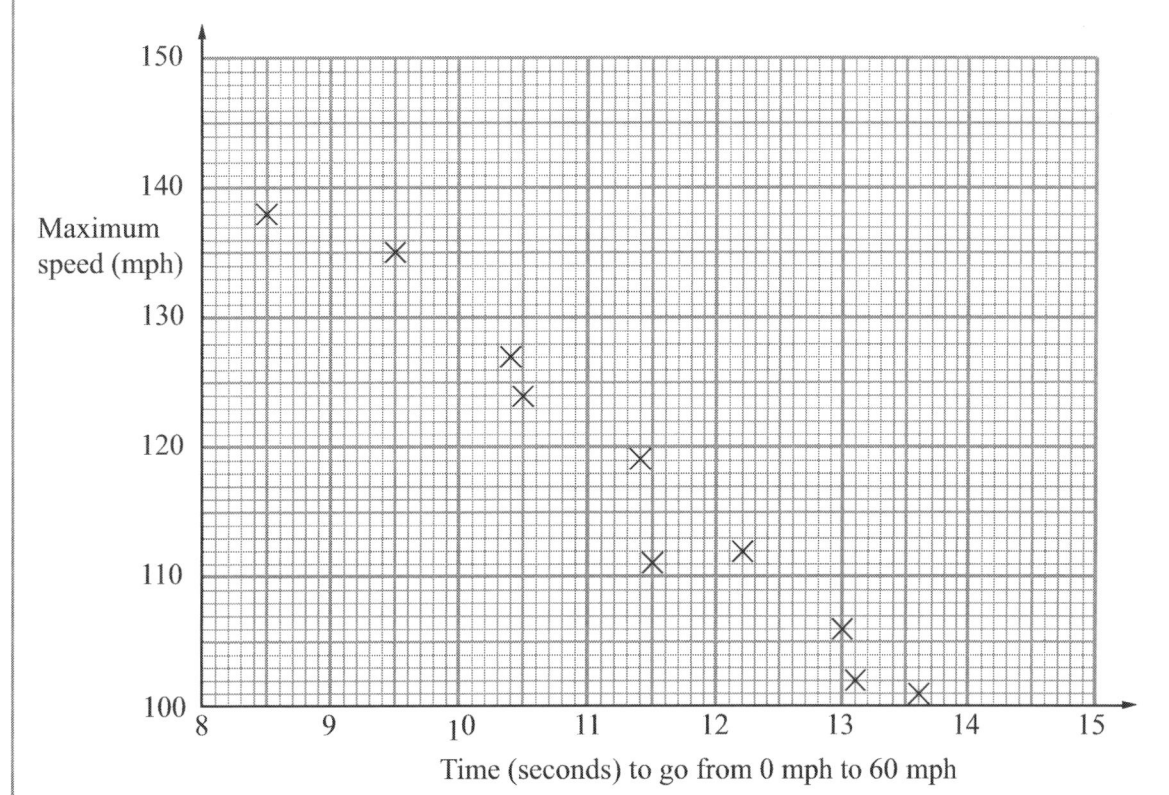

(a) What type of correlation does this scatter graph show?

....
(1)

The time a car takes to go from 0 mph to 60 mph is 11 seconds.

(b) Estimate the maximum speed for this car.

.... mph
(2)

(Total for Question 2 is 3 marks)

3 The scatter graph shows information about eight dogs.
It shows the height and the length of each dog.

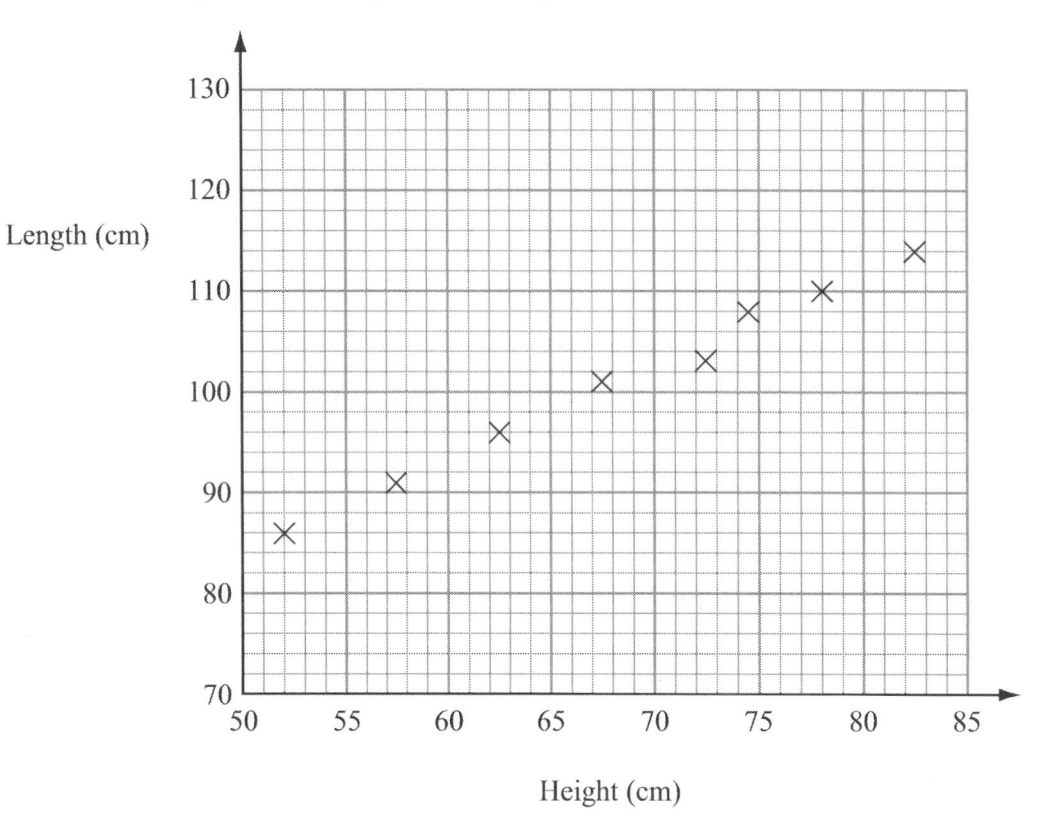

The table gives the height and the length of two more dog.

Height (cm)	65	80
Length (cm)	100	110

(a) On the scatter graph, plot the information from the table.

(1)

(b) Describe the relationship between the height and the length of these dog.

....

(1)

The height of a dog is 76 cm.

(c) Estimate the length of this dog.

.... cm

(2)

(Total for Question 3 is 4 marks)

4   Some students revised for a mathematics exam.
    They used a private tutor.
    The scatter graph shows the times seven students spent with the tutor and the marks the students got in the mathematics exam.

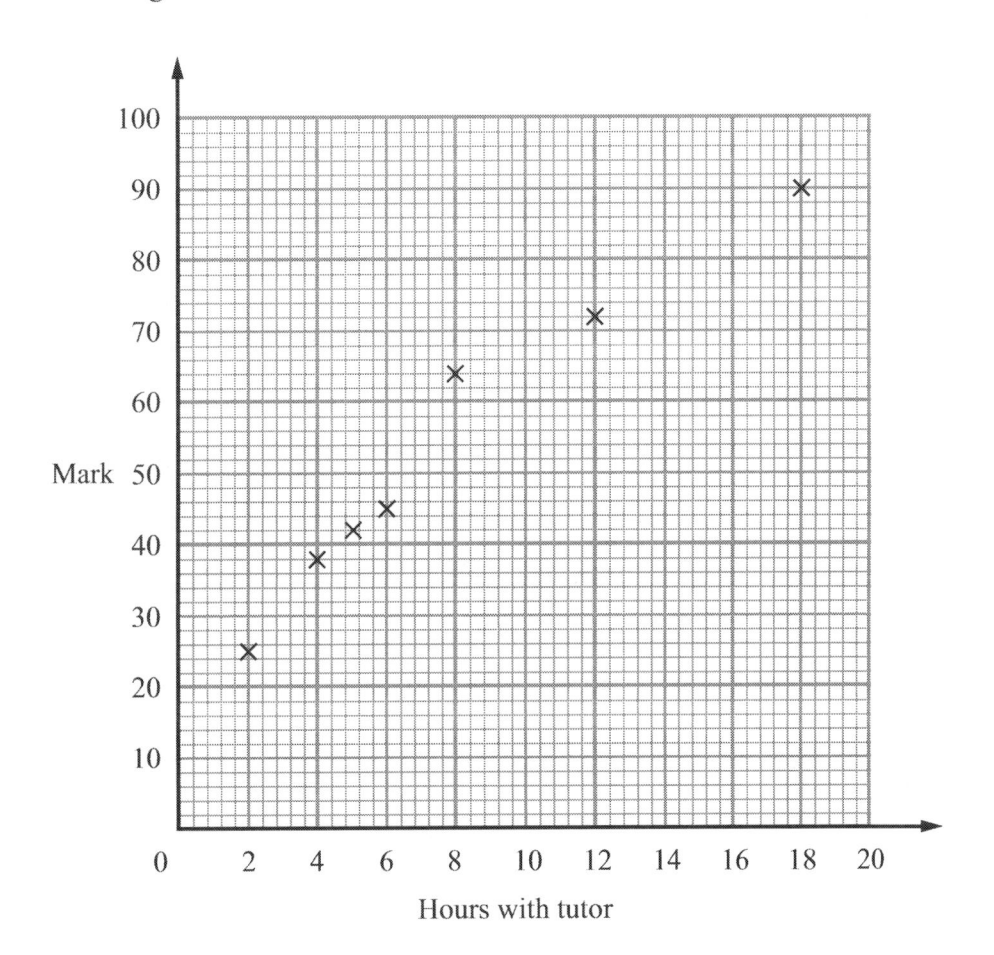

Here is the information for 3 more students.

Hours with tutor	7	10	16
Mark	50	56	78

(a) Plot this information on the scatter graph.

(1)

(b) What type of correlation does this scatter graph show?

....
(1)

(c) Draw a line of best fit on the scatter graph.
(1)

(Total for Question 4 is 3 marks)

5 The scatter graph shows information for some weather stations.
It shows the height of each weather station above sea level (m) and the mean August midday temperature (°C) for that weather station.

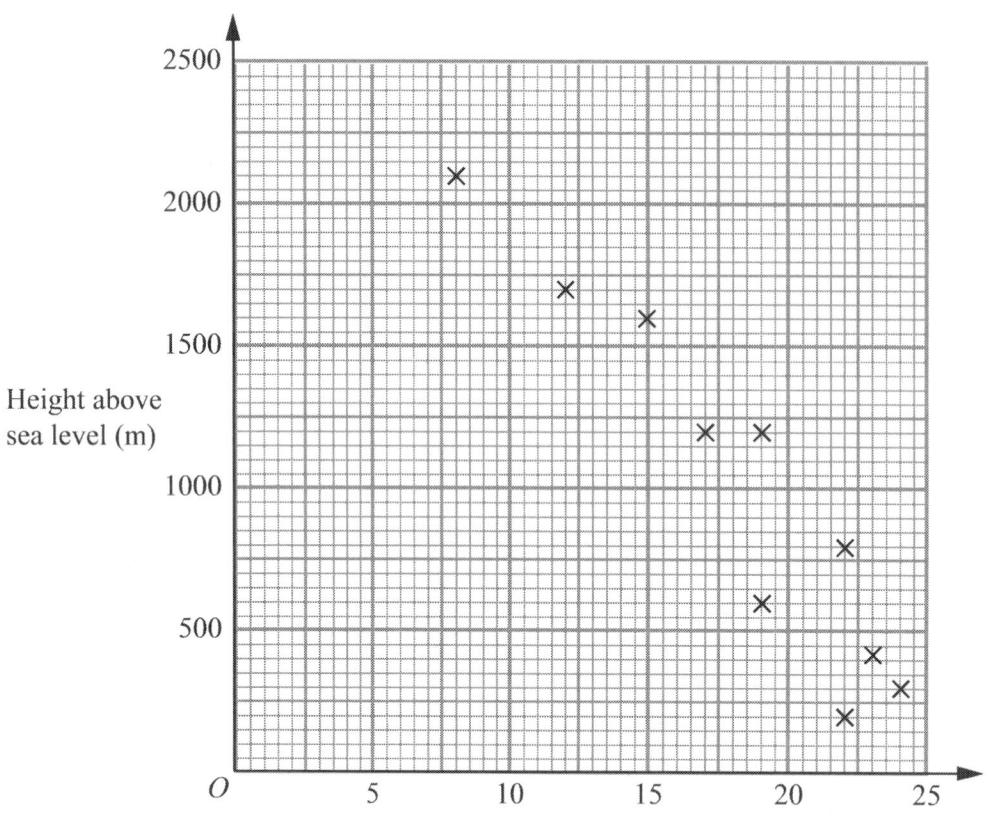

The table shows this information for two more weather stations.

Height of weather station above sea level (m)	1000	500
Mean August midday temperature (°C)	20	22

(a) Plot this information on the scatter graph.

(1)

(b) What type of correlation does this scatter graph show?

....

(1)

(c) Draw a line of best fit on the scatter graph.

(1)

(Total for Question 5 is 3 marks)

6   Mr Davies sells umbrellas.

The scatter graph shows some information about the number of umbrellas he sold and the rainfall, in cm, each month last year.

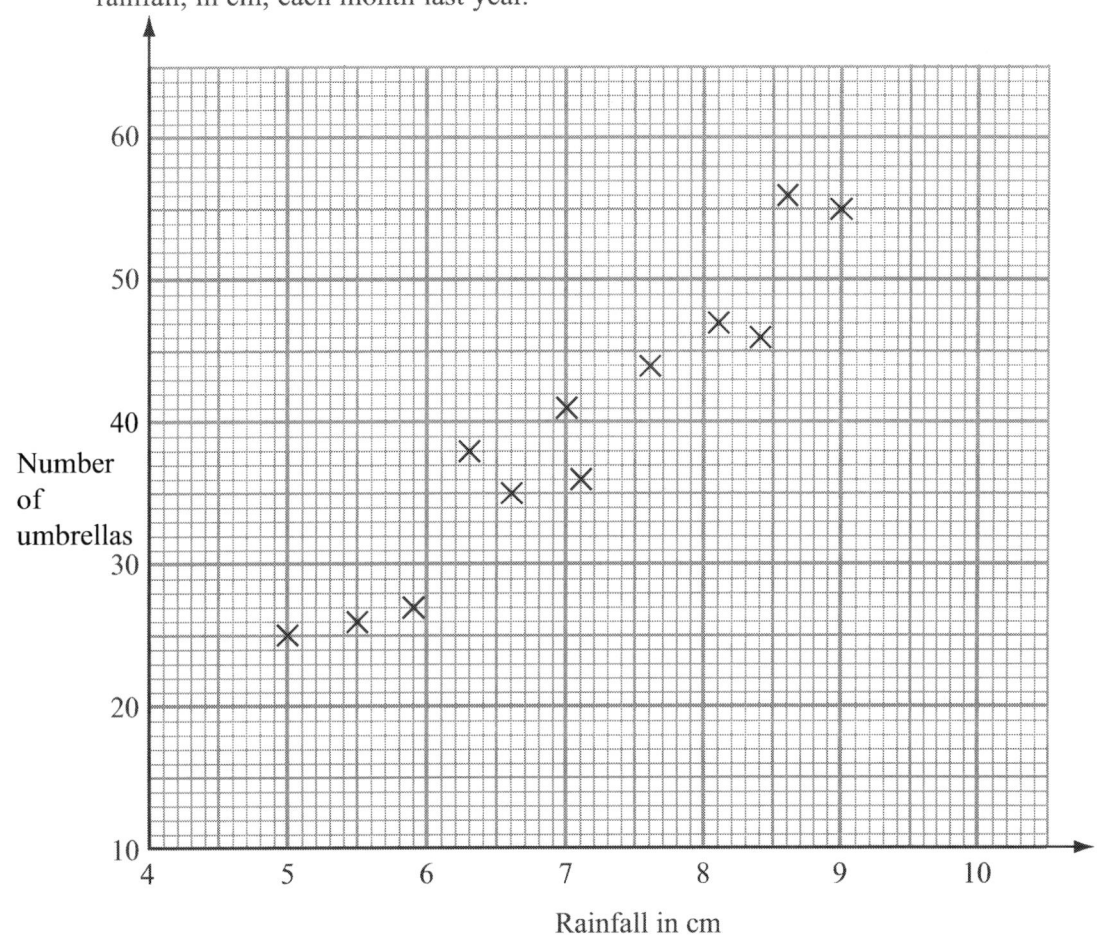

In January of this year, the rainfall was 6.1 cm.

During January, Mr Davies sold 33 umbrellas.

(a) Show this information on the scatter graph.
(1)

(b) What type of correlation does this scatter graph show?

....
(1)

In February of this year, Mr Davies sold 39 umbrellas.

(c) Estimate the rainfall for February.

.... cm
(2)

(Total for Question 6 is 4 marks)

7 Sophie reads eight books.

For each book she recorded the number of pages and the time she takes to read it.

The scatter graph shows information about her results.

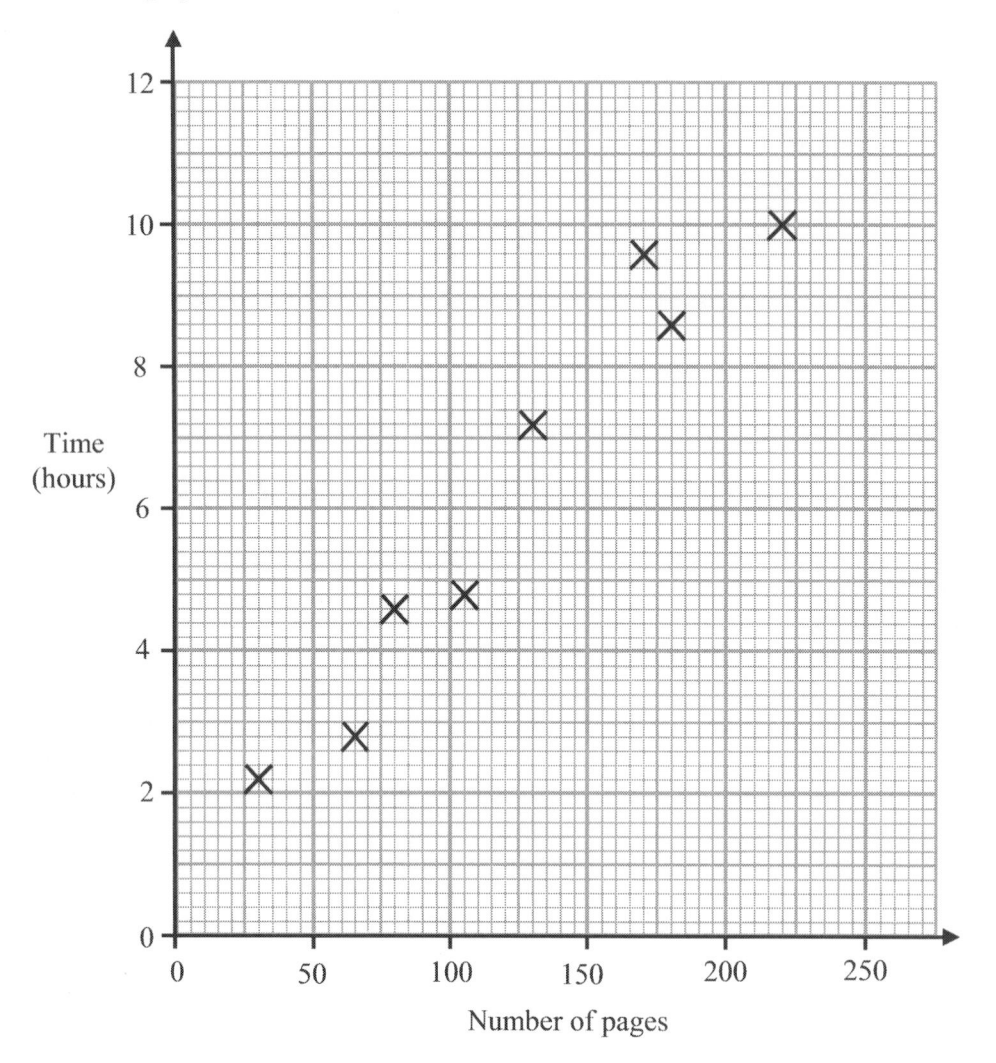

(a) Describe the relationship between the number of pages in a book and the time Sophie takes to read it.

....

(1)

Sophie reads another book.
The book has 200 pages.

(b) Estimate the time it takes Sophie to read it.

.... hours

(2)

(Total for Question 7 is 3 marks)

8 The table shows the cost and length of different tram journeys across a city.

Length of journey (miles)	1.8	2.1	2.2	2.5	3.2	3.7	4.0	4.6	5.8	6.4
Cost of journey (£)	0.90	0.80	1.50	1.60	2.00	2.20	2.40	2.90	3.10	3.40

(a) Draw a scatter diagram for the data on the grid below.

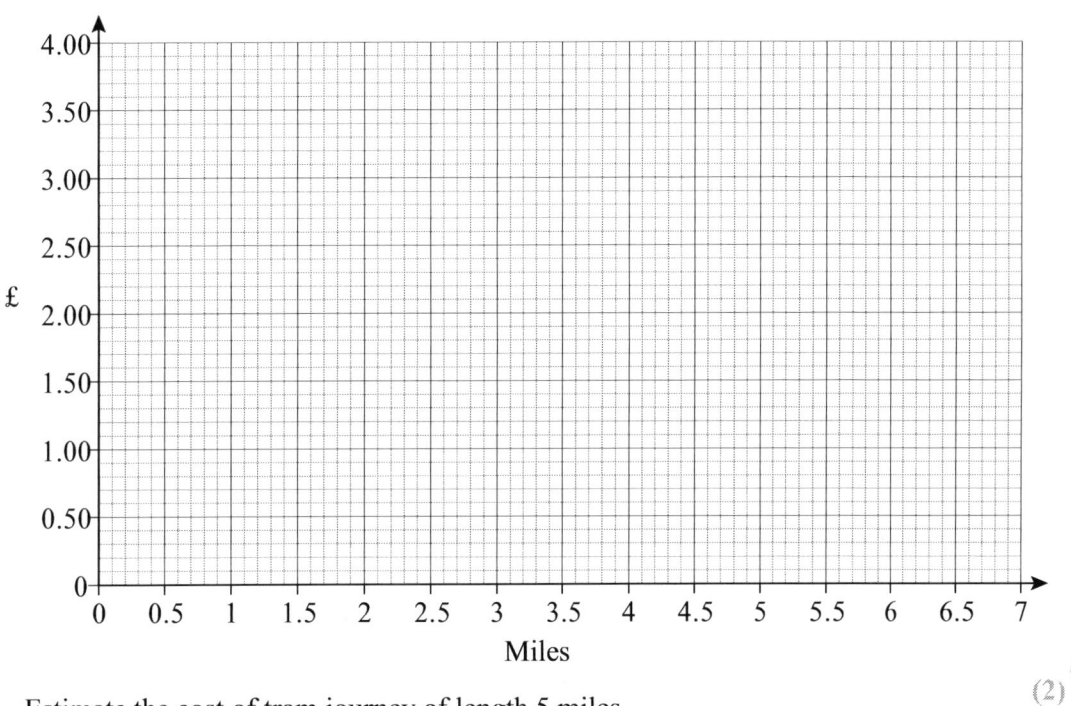

(2)

(b) Estimate the cost of tram journey of length 5 miles.
Give your answer to the nearest ten pence.

£ ....

(2)

(Total for Question 8 is 4 marks)

9 In a survey, the outside temperature and the number of units of electricity used for heating were recorded for ten homes.

The scatter diagram shows this information.

Sheila says,

"On average the number of units of electricity used for heating decreases by 4 units for each °C increase in outside temperature."

(a) Is Sheila right?
Show how you get your answer.

....

....

(3)

(b) You should **not** use a line of best fit to predict the number of units of electricity used for heating when the outside temperature is 30 °C.

Give one reason why.

....

(1)

(Total for Question 9 is 4 marks)

Write your name here

Surname

Other names

In the style of:
Pearson Edexcel
Level 1/Level 2 GCSE (9 - 1)

Centre Number

Candidate Number

Mathematics

Sequences

Higher Tier

GCSE style questions arranged by topic

Paper Reference
1MA1/3H

You must have: Ruler graduated in centimetres and millimetres, protractor, pair of compasses, pen, HB pencil, eraser, calculator.

Total Marks

Instructions

- Use **black** ink or ball-point pen.
- **Fill in the boxes** at the top of this page with your name, centre number and candidate number.
- Answer **all** questions.
- Answer the questions in the spaces provided
  – there may be more space than you need.
- **Calculators may be used.**
- If your calculator does not have a π button, take the value of π to be 3.142 unless the question instructs otherwise.
- Diagrams are **NOT** accurately drawn, unless otherwise indicated.
- You must **show all your working out**.

Information

- The total mark for this paper is
- The marks for **each** question are shown in brackets
  – use this as a guide as to how much time to spend on each question.

Advice

- Read each question carefully before you start to answer it.
- Keep an eye on the time.
- Try to answer every question.
- Check your answers if you have time at the end.

Turn over ▶

© Peter Bland

1   Here are some patterns made from squares.

    Pattern number 1   Pattern number 2   Pattern number 3

    (a) The diagram below shows part of Pattern number 5
        Complete the diagram for Pattern number 5

        Pattern number 4   Pattern number 5

    (1)

    (b) Complete the table.

Pattern number	1	2	3		
Number of squares	5	9	13		

    (1)

    (c) Find the number of squares used for Pattern number 12

    ....
    (1)

    (Total for Question 1 is 3 marks)

2 Here are some patterns made using sticks.

Pattern number 1   Pattern number 2   Pattern number 3

(a) In the space below, complete Pattern number 4.

　　　　　　Pattern number 4

(1)

(b) Complete the table.

Pattern number	1	2			
Number of sticks	4	7	10		

(1)

(c) How many sticks are used in Pattern number 10?

....

(1)

(Total for Question 2 is 3 marks)

3 Here are some patterns made with dots.

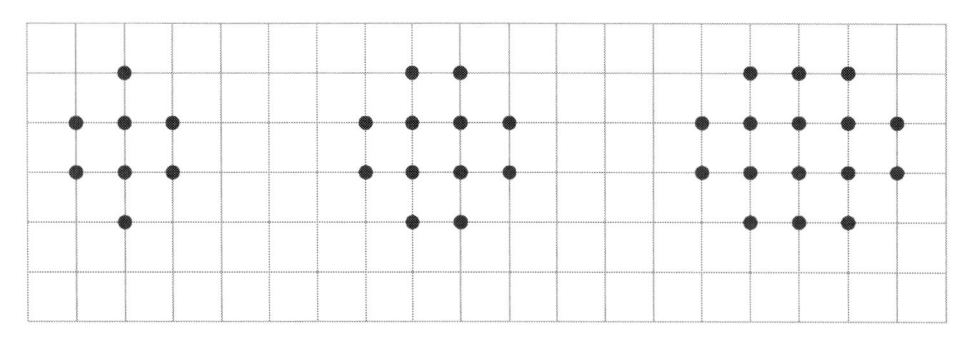

Pattern number 1   Pattern number 2   Pattern number 3

(a) In the space below, draw Pattern number 4

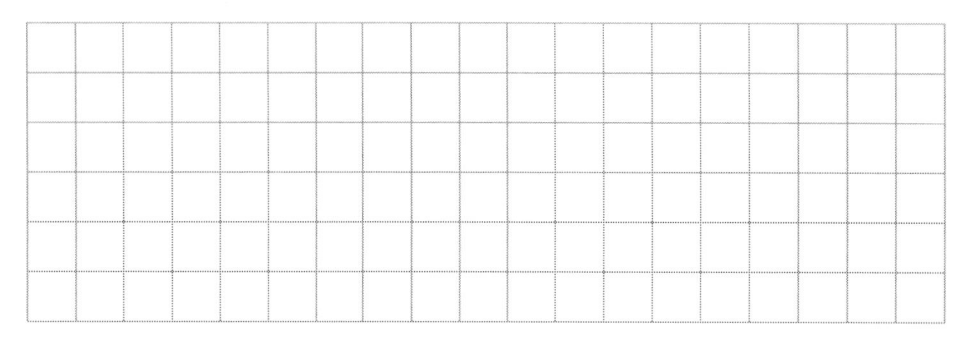

Pattern number 4

(1)

(b) Complete the table.

Pattern number	1	2			
Number of dots	8	12	16		

(2)

(Total for Question 3 is 3 marks)

4   The first even number is 2

   (a) Write down the 4th even number.

   ....
   (1)

   Here are some patterns made from sticks.

   Pattern number 1   Pattern number 3

   (b) Draw Pattern number 4

   Pattern number 4
   (1)

   (c) Complete the table.

Pattern number	1	2			
Number of sticks	3	6	9		

   (2)

   Jenny wants to find the number of sticks in Pattern number 100

   (d) Write down a method she could use.

   ....
   ....
   (1)

   (Total for Question 4 is 5 marks)

5 Here is a sequence of patterns made from grey squares and white squares.

Pattern Number 1

Pattern Number 2

Pattern Number 3

(a) Complete Pattern Number 5

Pattern Number 4

Pattern Number 5

(1)

(b) Complete the table.

Pattern Number	1	2			
Total number of squares	3	6			

(1)

One of the patterns in the sequence has 10 grey squares.

(c) How many white squares does this pattern have?

....

(1)

Another pattern in the sequence has a total of 18 squares.

(d) How many grey squares does the pattern have?

....

(1)

(Total for Question 5 is 4 marks)

6 Here are the first four terms of a number sequence.

5   9   13   17

(a) (i) Write down the next term of the number sequence.

..........21..........

(ii) Explain how you found your answer.

term-to-term rule is +4

(2)

The 24th term of the number sequence is 97

(b) Work out the 25th term of the number sequence.

..........101..........

(3)

(Total for Question 6 is 3 marks)

7   The *n*th term of a number sequence is given by $3n+1$

(a) Work out the first **three** terms of the number sequence.

....

(1)

Here are the first four terms of another number sequence.

   1   5   9   13

(b) Find, in terms of *n*, an expression for the *n*th term of this number sequence.

...............4n − 3...............

(2)

(Total for Question 7 is 3 marks)

8 Write down the next term in each sequence.

(a)(i) 5 8 11 14 ....

....
(1)

(a)(ii) 6 4 2 0 ....

....
(1)

(a)(iii) 2 4 8 16 ....

....
(1)

(b) The numbers in this sequence increase by the same amount each time.

11 35

What are the missing numbers?

....
(1)

(Total for Question 8 is 4 marks)

9   The *n*th term of a sequence is $100 - 3n$.

(a) Work out the first three terms.

..........,..........,..........
(2)

(b) Work out the first term of the sequence that is negative.

....
(2)

(Total for Question 9 is 4 marks)

10 (a) Here are the first three terms of a sequence.

12   8   6   ....

The rule for working out the next term in the sequence is

| Add 4 to the previous term and then divide by 2 |

Work out the first term that is **not** a whole number.

....

(2)

(b) This sequence uses the same rule.

| Add 4 to the previous term and then divide by 2 |

The third term of this sequence is 9.

....   9   ....

Work out the first term.

....

(3)

(Total for Question 10 is 5 marks)

11 (a) Write down the next term of each sequence.

(a) (i)   3   8   13   18   ....

....

(1)

(a) (ii)   5.1   5.3   5.5   5.7   ....

....

(1)

(a) (iii)   2   −1   −4   −7   ....

....

(1)

(b) Here is a different sequence.
The third term is 20 and the fourth term is 36.

     ……   ……   20   36   ……

The term to term rule for this sequence is

Double and subtract four

Work out the first term of the sequence.

....

(2)

(Total for Question 11 is 5 marks)

12 (a) The numbers in this sequence decrease by the same amount each time.

        74   58   50   42   ......

What are the **two** missing numbers?

.... and ....

(2)

(b) The numbers in this different sequence decrease by the same amount each time.

        26   6

What are the **three** missing numbers?

.... , ... , ...

(2)

(Total for Question 12 is 4 marks)

13 (a) Here are the first two terms of a sequence.

5   4   ....

The rule for finding the next term in the sequence is

| Multiply the previous term by 2 and subtract 6 |

Work out the first negative term of the sequence.

(2)

(b) Here are the first three terms of another sequence.

1   4   7   ...   .

Which of the following is the nth term for this sequence? Circle the correct answer.

$n + 3$ $3n + 1$ $3n - 2$ $3n + 2$

(1)

(Total for Question 13 is 3 marks)

14 (a)   A sequence starts

  49   46   43   40

(a) (i)   Write down the next two terms.

.... and ....

(2)

(a) (ii)   What is the rule for continuing the sequence?

....

(1)

(b)   Another sequence starts

  57   50   43   36

This sequence is continued.

What is the first negative number in this sequence?

....

(1)

(c)   The first sequence is also continued.

The two sequences have the number 43 in common.

What is the next number that the two sequences have in common?

....

(2)

(Total for Question 14 is 6 marks)

15 Here are the first four terms of an arithmetic sequence.

      6   10   14   18

(a) Write an expression, in terms of n, for the nth term of this sequence.

....
(2)

The nth term of a different arithmetic sequence is $3n + 5$

(b) Is 108 a term of this sequence?
Show how you get your answer.

(2)

(Total for Question 15 is 4 marks)

16 (a) The nth term of a sequence is $2^n + 2^{n-1}$

Work out the 10th term of the sequence.

....
(1)

(b) The nth term of a different sequence is $4(2^n + 2^{n-1})$

Circle the expression that is equivalent to $4(2^n + 2^{n-1})$

$2^{n+2} + 2^{n+1}$ $\qquad\qquad$ $2^{2n} + 2^{2(n-1)}$

$8^n + 8^{n-1}$ $\qquad\qquad$ $2^{n+2} + 2^{n-1}$

(1)

(Total for Question 16 is 2 marks)

17 **(a)** Find the *n*th term of this linear sequence.

$$8 \quad 11 \quad 14 \quad 17$$

(a) ....

(2)

(b) Here is a quadratic sequence.

$$2 \quad 14 \quad 36 \quad 68$$

The expression for the *n*th term of this sequence is $pn^2 + qn$.

Find the value of *p* and the value of *q*.

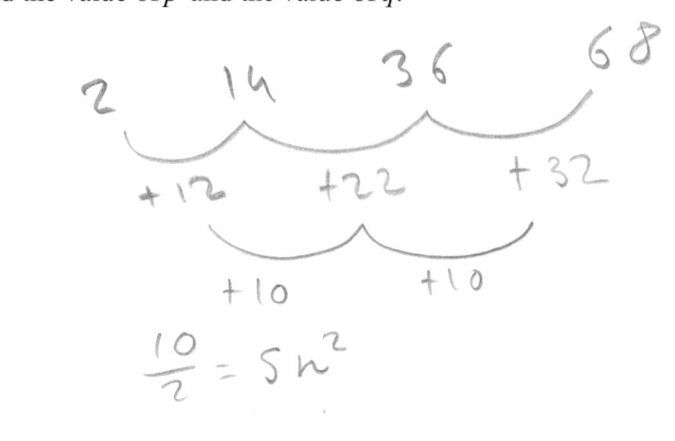

$$2 \quad 14 \quad 36 \quad 68$$

+12 +22 +32

+10 +10

$\frac{10}{2} = 5n^2$

$5n^2 = \quad 5 \quad 20 \quad 45 \quad 80$
$ 2 \quad 14 \quad 36 \quad 68$
$\overline{}$
$ 3 \quad 6 \quad 9 \quad 12$

+3 +3 +3

(b) *p* = ...5.........

q = ...3.........

(4)

(Total for Question 17 is 6 marks)

18 Here are the first three patterns in a sequence.
The patterns are made from triangles and rectangles.

pattern number 1   pattern number 2   pattern number 3

(a) How many triangles are there in pattern number 7?

....
(2)

Hassan says

"There are 4 rectangles in pattern number 3 so there will be 8 rectangles in pattern number 6"

(b) Is Hassan right?

Give a reason for your answer.

(1)

(Total for Question 18 is 3 marks)

Write your name here

Surname

Other names

In the style of:
Pearson Edexcel
Level 1/Level 2 GCSE (9 - 1)

Centre Number

Candidate Number

Mathematics
Simultaneous Equations
Higher Tier

GCSE style questions arranged by topic

Paper Reference
1MA1/2H

You must have: Ruler graduated in centimetres and millimetres, protractor, pair of compasses, pen, HB pencil, eraser, calculator.

Total Marks

Instructions

- Use **black** ink or ball-point pen.
- **Fill in the boxes** at the top of this page with your name, centre number and candidate number.
- Answer **all** questions.
- Answer the questions in the spaces provided
  – *there may be more space than you need.*
- **Calculators may be used.**
- If your calculator does not have a π button, take the value of π to be 3.142 unless the question instructs otherwise.
- Diagrams are **NOT** accurately drawn, unless otherwise indicated.
- You must **show all your working out**.

Information

- The total mark for this paper is
- The marks for **each** question are shown in brackets
  – *use this as a guide as to how much time to spend on each question.*

Advice

- Read each question carefully before you start to answer it.
- Keep an eye on the time.
- Try to answer every question.
- Check your answers if you have time at the end.

Turn over ▶

© Peter Bland

1 Solve the simultaneous equations

$3x + 2y = 8$
$2x + 5y = -2$

$3x + 2y = 8$ ① ×2
$2x + 5y = -2$ ② ×3

$6x + 4y = 16$
$-\ 6x + 15y = -6$
_____
$11y =$

$x = $ ....

$y = $ ....

(Total for Question 1 is 4 marks)

2 Solve the simultaneous equations

$6x + 2y = -3$
$4x - 3y = 11$

$x = $ ...., $y = $ ....

(Total for Question 2 is 4 marks)

3 Solve the simultaneous equations

$x^2 + y^2 = 5$
$y = 3x + 1$

$x^2 + (3x+1)(3x+1) = 5$

$x^2 + 9x^2 + 6x + 1 = 5$

$\dfrac{10x^2 + 6x - 4 = 0}{2}$

$5x^2 + 3x - 2 = 0$

$5x^2 + 5x - 2x - 2 = 0$

$5x(x+1) - 2(x+1) = 0$

$(x+1)(5x-2) = 0$

$x = -1$ or $x = \dfrac{2}{5}$

$5x - 2 = 0$
$5x = 2$
$x = \dfrac{2}{5}$

sub $x = -1$ into $y = 3x+1$
$y = 3(-1) + 1$
$y = -3 + 1$
$y = -2$

or sub $x = \dfrac{2}{5}$ into $y = 3x+1$

$y = \dfrac{3}{1} \times \dfrac{2}{5} + 1$

$y = \dfrac{6}{5} + 1$

$y = \dfrac{11}{5}$

$x = -1$ $y = -2$
or $x = 2/5$ $y = 11/5$

(Total for Question 3 is 6 marks)

4 Solve the simultaneous equations

   $4x + y = -1$
   $4x - 3y = 7$

   $x = $ .... $y = $ ....

   (Total for Question 4 is 3 marks)

5

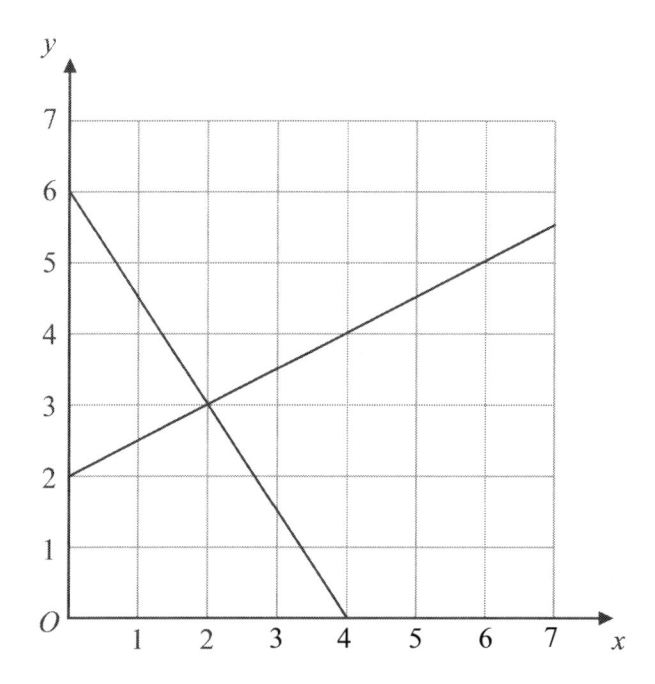

The diagram shows graphs of $y = \dfrac{1}{2}x + 2$

and $2y + 3x = 12$

(a) Use the diagram to solve the simultaneous equations

$$y = \dfrac{1}{2}x + 2$$

$$2y + 3x = 12$$

x = ... y = ...

(1)

(b) Find an equation of the straight line which is parallel to the line $y = \dfrac{1}{2}x + 2$ and passes through the point (0, 4).

....

(2)

(Total for Question 5 is 3 marks)

6 Solve the simultaneous equations

$$6x + 2y = -3$$
$$4x - 3y = 11$$

$x = $ ...., $y = $ ....

(Total for Question 6 is 4 marks)

7 Solve the simultaneous equations
$$4x + y = 10$$
$$2x - 3y = 19$$

$x = $ ....

$y = $ ....

(Total for Question 7 is 3 marks)

8 Solve algebraically the simultaneous equations

$$x^2 + y^2 = 25$$
$$y - 2x = 5$$

$(y = 2x + 5)$

$x^2 + (2x+5)^2 = 25$

$x^2 + (2x+5)(2x+5) = 25$

$x^2 + (2x+5)(2x+5) - 25 = 0$

$x^2 + 4x^2 + 10x + 10x + 25 - 25 = 0$

$5x^2 + 20x = 0$

$5x(x+4) = 0$

$x = -4$
or
$x = 0$

$y = 5$
$y = -3$

(Total for Question 8 is 3 marks)

9   Here is the graph of $4x - 3y = 12$ for values of x from 0 to 4

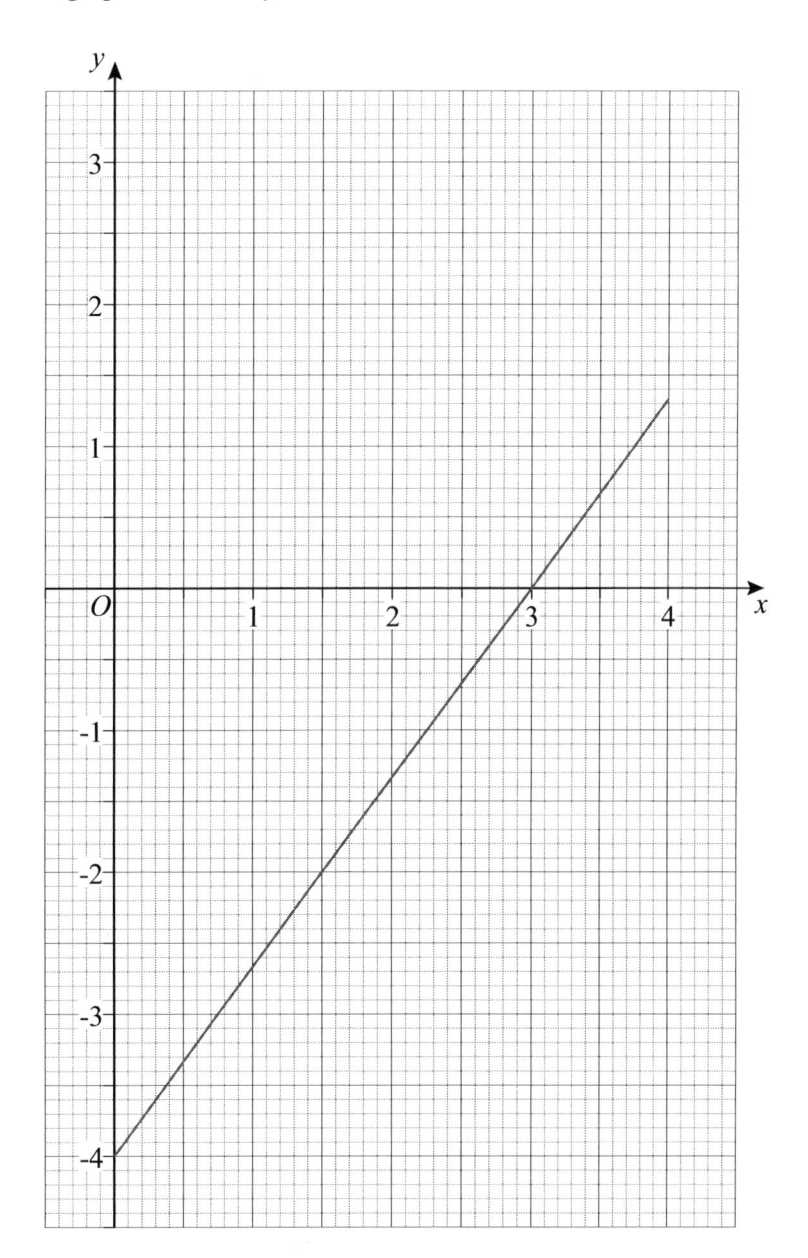

By drawing a second graph on the grid,

work out an approximate solution to the simultaneous equations

$$4x - 3y = 12 \quad \text{and} \quad 3x + 2y = 6$$

....

(Total for Question 9 is 3 marks)

Solve these simultaneous equations algebraically.

$$y = x - 3$$
$$y = 2x^2 + 8x - 7$$

$x - 3 = 2x^2 + 8x - 7$

$x = 2x^2 + 8x - 4$

$2x^2 + 7x - 4 = 0$

$x = $ .... , $y = $ ....
$x = $ .... , $y = $ ....

(Total for Question 10 is 6 marks)

Write your name here

Surname

Other names

In the style of:
**Pearson Edexcel
Level 1/Level 2 GCSE (9 - 1)**

Centre Number

Candidate Number

Mathematics
Surds and Indices

Higher Tier

GCSE style questions arranged by topic

Paper Reference
1MA1/1H

You must have: Ruler graduated in centimetres and millimetres, protractor, pair of compasses, pen, HB pencil, eraser.

Total Marks

Instructions

- Use **black** ink or ball-point pen.
- **Fill in the boxes** at the top of this page with your name, centre number and candidate number.
- Answer **all** questions.
- Answer the questions in the spaces provided
  – *there may be more space than you need.*
- **Calculators may not be used.**
- Diagrams are **NOT** accurately drawn, unless otherwise indicated.
- You must **show all your working out**.

Information

- The total mark for this paper is
- The marks for **each** question are shown in brackets
  – *use this as a guide as to how much time to spend on each question.*

Advice

- Read each question carefully before you start to answer it.
- Keep an eye on the time.
- Try to answer every question.
- Check your answers if you have time at the end.

Turn over ▶

© Peter Bland

1 Work out $(2 + \sqrt{5})(2 - \sqrt{5})$

Give your answer in its simplest form.

$(2+\sqrt{5})(2-\sqrt{5})$
$4 - 2\sqrt{5} + 2\sqrt{5} - 5$
$4 - 5 = -1$

..........−1..........
(1)

(Total for Question 1 is 1 mark)

2 (a) Write down the value of $64^{\frac{1}{2}}$ $= \sqrt{64} = 8$

..........8..........
(1)

(b) Write $\sqrt{45}$ in the form $k\sqrt{5}$, where k is an integer.

$\sqrt{9 \times 5}$
$3\sqrt{5}$

..........$3\sqrt{5}$..........
(2)

(Total for Question 2 is 3 marks)

3 Find the value of

(i) 8^0

........1........ (1)

(ii) $64^{\frac{1}{2}}$

........8........ (1)

(iii) $\left(\frac{27}{8}\right)^{\frac{2}{3}}$ $\quad \dfrac{27^{\frac{2}{3}}}{8^{\frac{2}{3}}}$

.... (2)

(Total for Question 3 is 4 marks)

4 (a) Simplify $4x \times 5y$

20xy (1)

(b) Simplify $x \times x \times x \times x$

x^4 (1)

(c) Expand $4(3n - 7)$

$12n - 28$

$12n - 28$ (2)

(d) Expand and simplify $2(2x + 3) + 3(x + 1)$

$4x + 6 + 3x + 3$
$7x + 9$

$7x + 9$ (2)

(e) Simplify $n^2 \times n$

n^3 (1)

(f) Simplify $p^5 \div p^3$

p^2 (1)

(Total for Question 4 is 8 marks)

5 (a) Simplify $q^5 \times q^4$

q^9 (1)

(b) Simplify $r^5 \div r^2$

r^3 (1)

(c) Simplify $12tv^6 \div 6tv^5$

$2tv^1$ (2)

(d) Simplify $(9w^2y^6)^{\frac{1}{2}}$

$4.5wy^3$

$4.5wy^3$ (2)

(e) For $y > 1$, write the following expressions in order of size.
Start with the expression with the least value.

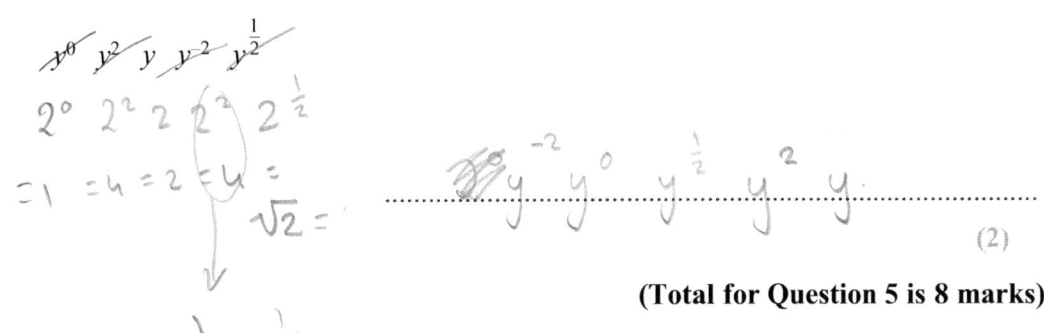

$y^0 \quad y^{-2} \quad y \quad y^2 \quad y^{\frac{1}{2}}$

$2^0 \quad 2^{-2} \quad 2 \quad 2^2 \quad 2^{\frac{1}{2}}$
$=1 \quad =4 \quad =2 \quad =4 \quad =\sqrt{2}$

$\frac{1}{2^2} = \frac{1}{4}$

$y^{-2} \quad y^0 \quad y^{\frac{1}{2}} \quad y^2 \quad y$ (2)

(Total for Question 5 is 8 marks)

6 (a) Simplify $n^3 \times n^4$

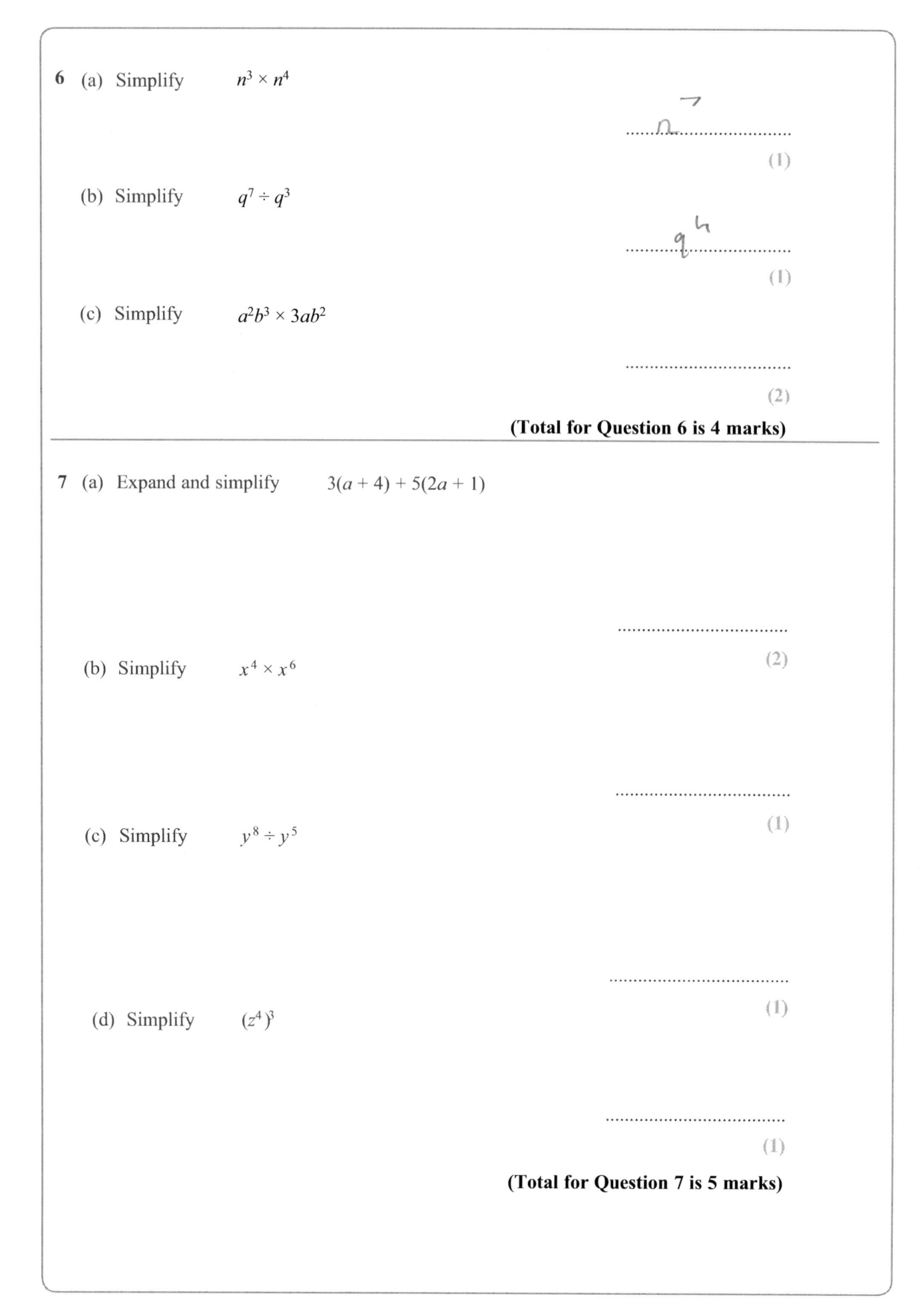

.... n^7 ....
(1)

(b) Simplify $q^7 \div q^3$

.... q^4 ....
(1)

(c) Simplify $a^2b^3 \times 3ab^2$

....
(2)

(Total for Question 6 is 4 marks)

7 (a) Expand and simplify $3(a + 4) + 5(2a + 1)$

....
(2)

(b) Simplify $x^4 \times x^6$

....
(1)

(c) Simplify $y^8 \div y^5$

....
(1)

(d) Simplify $(z^4)^3$

....
(1)

(Total for Question 7 is 5 marks)

8 (a) Simplify $v^6 \times v^2$

....
(1)

(b) Simplify $\dfrac{m^8}{m^3}$

....
(1)

(c) Simplify $(2y)^3$

....
(2)

(d) Simplify $3a^2h \times 4a^5h^4$

....
(2)

(Total for Question 8 is 6 marks)

9 Work out the value of $(9 \times 10^{-4}) \times (3 \times 10^{7})$
Give your answer in standard form.

....
(Total for Question 9 is 2 marks)

10 (a) Write down the value of $64^{\frac{1}{2}}$

............8.................
(1)

(b) Find the value of $\left(\dfrac{8}{125}\right)^{-\frac{2}{3}}$

....
(2)

(Total for Question 10 is 3 marks)

11 One uranium atom has a mass of 3.95×10^{-22} grams.

(a) Work out an estimate for the number of uranium atoms in 1 kg of uranium.

1 kg = 1000 g

.... (3)

(b) Is your answer to (a) an underestimate or an overestimate?
Give a reason for your answer.

....

....
(1)

(Total for Question 11 is 4 marks)

12 Write 0.000068 in standard form.

6.8×10^{-5}

.... 6.8×10^{-5} ....

(Total for Question 12 is 1 mark)

13 $a \times 10^4 + a \times 10^2 = 24\,240$ where a is a number.

Work out $a \times 10^4 - a \times 10^2$

Give your answer in standard form.

....
(2)

(Total for Question 13 is 2 marks)

14 Rationalise the denominator and simplify $\dfrac{10}{3\sqrt{5}}$

$\dfrac{10}{3\sqrt{5}} \times \dfrac{3\sqrt{5}}{3\sqrt{5}}$

$= \dfrac{10 \times (3\sqrt{5})}{(3\sqrt{5})(3\sqrt{5})}$

$= \dfrac{30\sqrt{5}}{6 + 3\sqrt{5} + 3\sqrt{5} + 5}$

$\dfrac{5\sqrt{5}}{11+\sqrt{5}}$

(2)

$= \dfrac{30\sqrt{5}}{6 + 6\sqrt{5} + 5}$

(Total for Question 14 is 2 marks)

$= \dfrac{\cancel{30}^{5}\sqrt{5}}{11+\cancel{6}\sqrt{5}}$

$= \dfrac{5\sqrt{5}}{11+\sqrt{5}}$

15 (a) Show that $\sqrt{396}$ can be written as $6\sqrt{11}$.

(2)

(b) Without using a calculator, show that $\dfrac{4+2\sqrt{2}}{2-\sqrt{2}}$ can be simplified to $6+4\sqrt{2}$

$$\dfrac{4+2\sqrt{2}}{2-\sqrt{2}} \times \dfrac{2-\sqrt{2}}{2-\sqrt{2}}$$

$$\dfrac{(4+2\sqrt{2})(2-\sqrt{2})}{(2-\sqrt{2})(2-\sqrt{2})}$$

$2 - 2\sqrt{2} - 2\sqrt{2} + 2$
$= -4\sqrt{2}$

(6)

(Total for Question 15 is 8 marks)

$(4+2\sqrt{2})(2-\sqrt{2})$
$= 8 - 4\sqrt{2} + 4\sqrt{2} - 4$
$= 2$

$\dfrac{2}{-4\sqrt{2}}$

16 (a) Without using a calculator, show that $\sqrt{20} = 2\sqrt{5}$

(2)

(b) The point A is shown on the unit grid below.
The point B is $2\sqrt{5}$ units from A and lies on the intersection of two grid lines.

Mark **one** possible position for B.

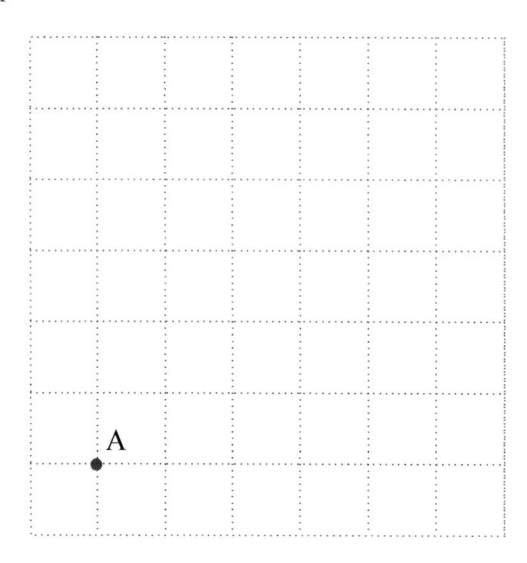

(3)

(Total for Question 16 is 5 marks)

17 The volume of Earth is 1.08×10^{12} km^3

The volume of Jupiter is 1.43×10^{15} km^3.

How many times larger is the radius of Jupiter than the radius of Earth?

Assume that Jupiter and Earth are both spheres.

Volume of sphere $= \dfrac{4}{3}\pi r^3$

....
(4)

(Total for Question 17 is 4 marks)

Write your name here

Surname

Other names

In the style of:
Pearson Edexcel
Level 1/Level 2 GCSE (9 - 1)

Centre Number

Candidate Number

Mathematics

Transformation of Curves

Higher Tier

GCSE style questions arranged by topic

Paper Reference
1MA1/1H

You must have: Ruler graduated in centimetres and millimetres, protractor, pair of compasses, pen, HB pencil, eraser.

Total Marks

Instructions

- Use **black** ink or ball-point pen.
- **Fill in the boxes** at the top of this page with your name, centre number and candidate number.
- Answer **all** questions.
- Answer the questions in the spaces provided
  – *there may be more space than you need.*
- **Calculators may not be used.**
- Diagrams are **NOT** accurately drawn, unless otherwise indicated.
- You must **show all your working out**.

Information

- The total mark for this paper is
- The marks for **each** question are shown in brackets
  – *use this as a guide as to how much time to spend on each question.*

Advice

- Read each question carefully before you start to answer it.
- Keep an eye on the time.
- Try to answer every question.
- Check your answers if you have time at the end.

Turn over

© Peter Bland

1

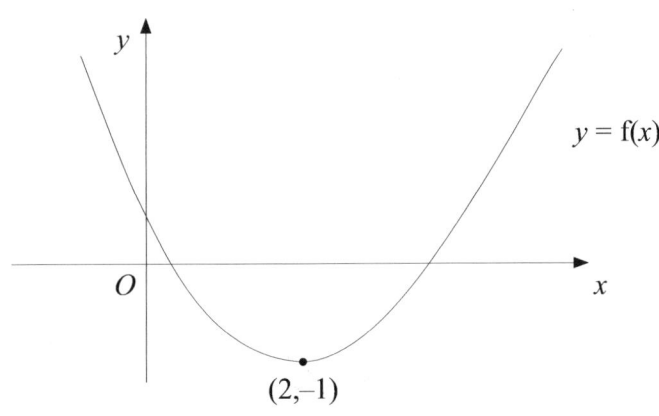

The diagram shows part of the curve with equation $y = f(x)$

The minimum point of the curve is at $(2, -1)$

(a) Write down the coordinates of the minimum point of the curve with equation

(i) $y = f(x - 2)$

$(4, -1)$

(ii) $y = 2f(x)$

$(2, -2)$

(iii) $y = f(2x)$

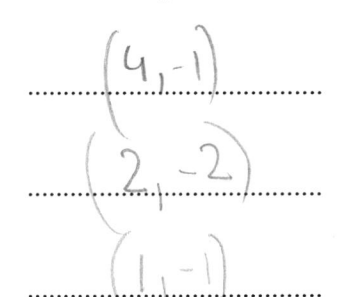

$(1, -1)$

(3)

The curve $y = f(x)$ is reflected in the y axis.

(b) Find the equation of the curve following this transformation.

$y = f(-x)$

(1)

The curve with equation $y = f(x)$ has been transformed to give the curve with equation $y = f(x) + 2$

(c) Describe the transformation.

The curve is translated by vector $\binom{0}{2}$

(1)

(Total for Question 1 is 5 marks)

2

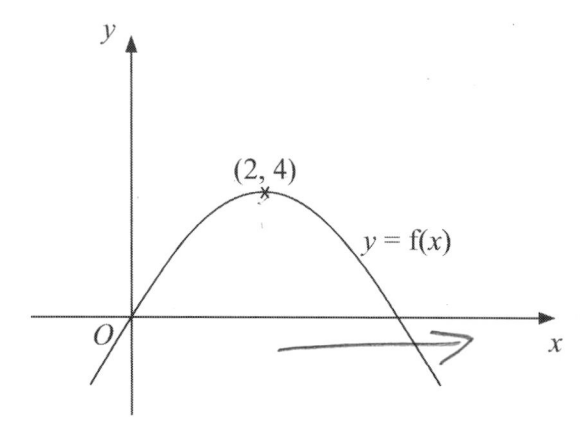

The diagram shows part of the curve with equation $y = f(x)$.

The coordinates of the maximum point of this curve are (2, 4).

Write down the coordinates of the maximum point of the curve with equation

(a) $y = f(x - 2)$

(....4...., ...4....)
(1)

(b) $y = 2f(x)$

(....2...., ...8....)
(1)

(Total for Question 2 is 2 marks)

3 The diagram shows a sketch of the curve $y = \sin x°$ for $0 \leqslant x \leqslant 360$

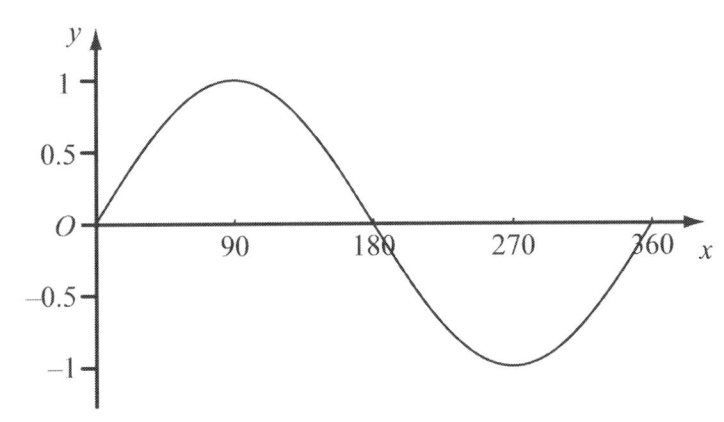

The exact value of $\sin 60° = \dfrac{\sqrt{3}}{2}$

(a) Write down the exact value of

(i) $\sin 120°$,

....

(ii) $\sin 300°$.

....
(2)

(b) On the grid below, sketch the graph of $y = 3\sin 2x°$ for $0 \leq x \leq 360$

(2)

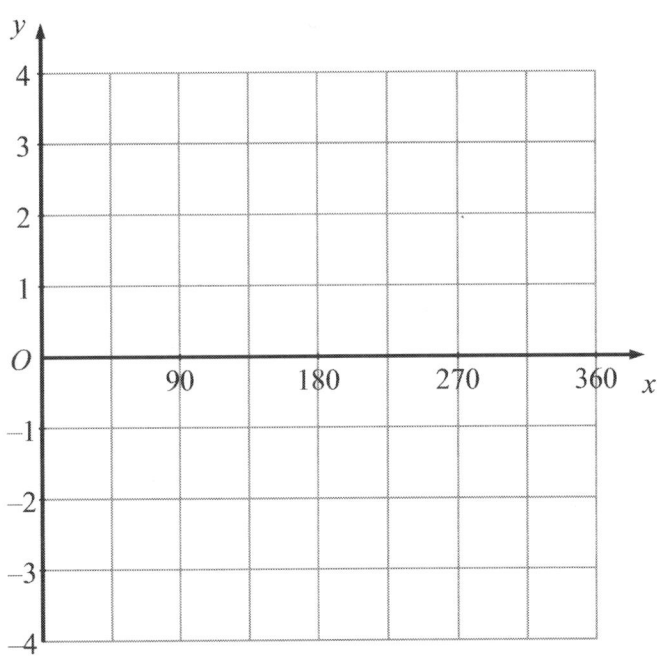

(Total for Question 3 is 4 marks)

4

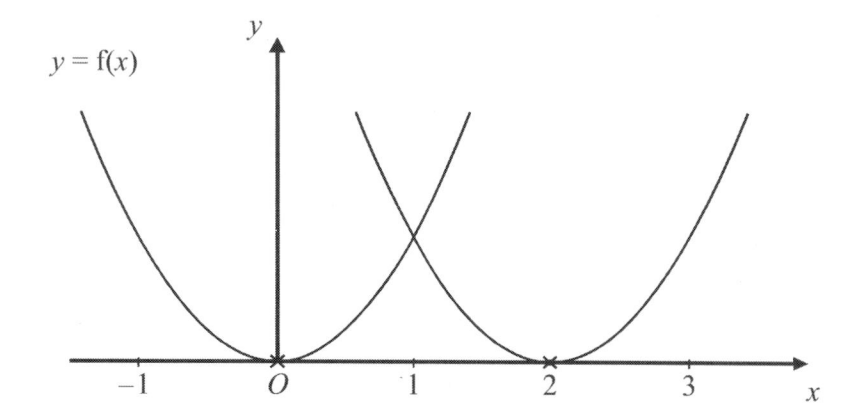

The curve with equation $y = f(x)$ is translated so that the point at $(0, 0)$ is mapped onto the point $(2, 0)$.

(a) Find an equation of the translated curve.

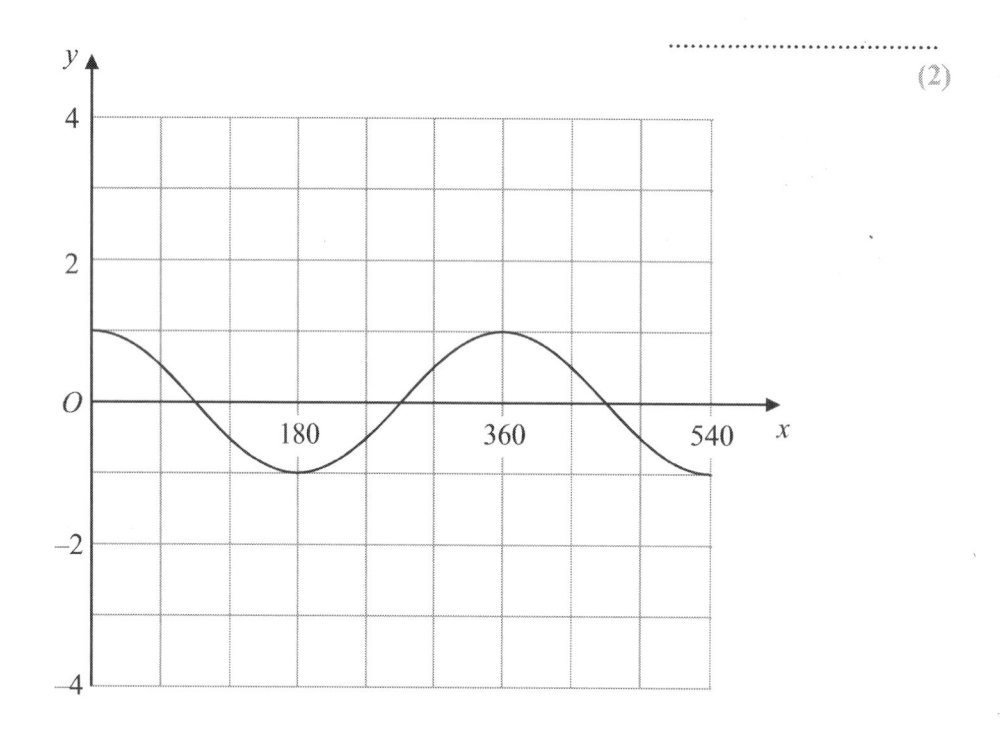

....
(2)

The grid shows the graph of $y = \cos x°$ for values of x from 0 to 540

(b) On the grid, sketch the graph of $y = 3\cos(2x°)$ for values of x from 0 to 540

(2)

(Total for Question 4 is 4 marks)

5 This is a sketch of the curve with the equation $y = f(x)$.
The only minimum point of the curve is at $P(3, -4)$.

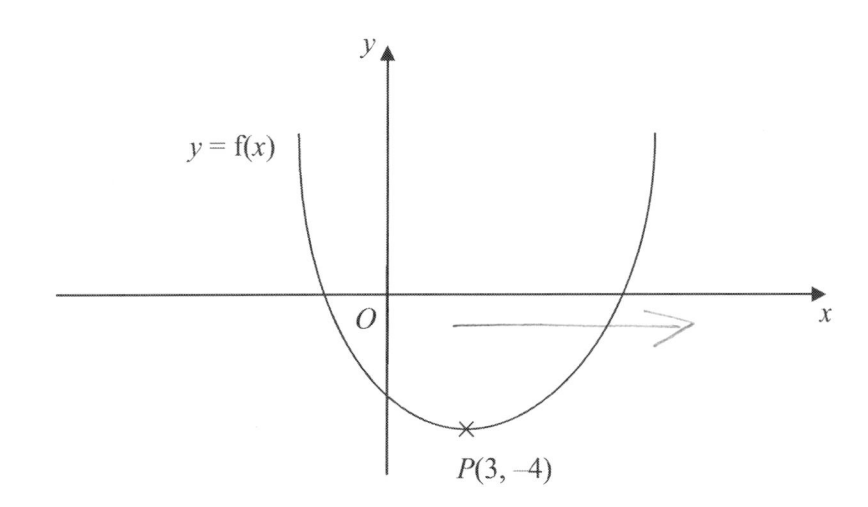

(a) Write down the coordinates of the minimum point of the curve with the equation $y = f(x - 2)$

(....5...., ...-4....)
(2)

(b) Write down the coordinates of the minimum point of the curve with the equation $y = f(x + 5) + 6$

(....-2...., ...2....)
(2)

(Total for Question 5 is 4 marks)

6 The graph of y = f(x) is shown on the grid.

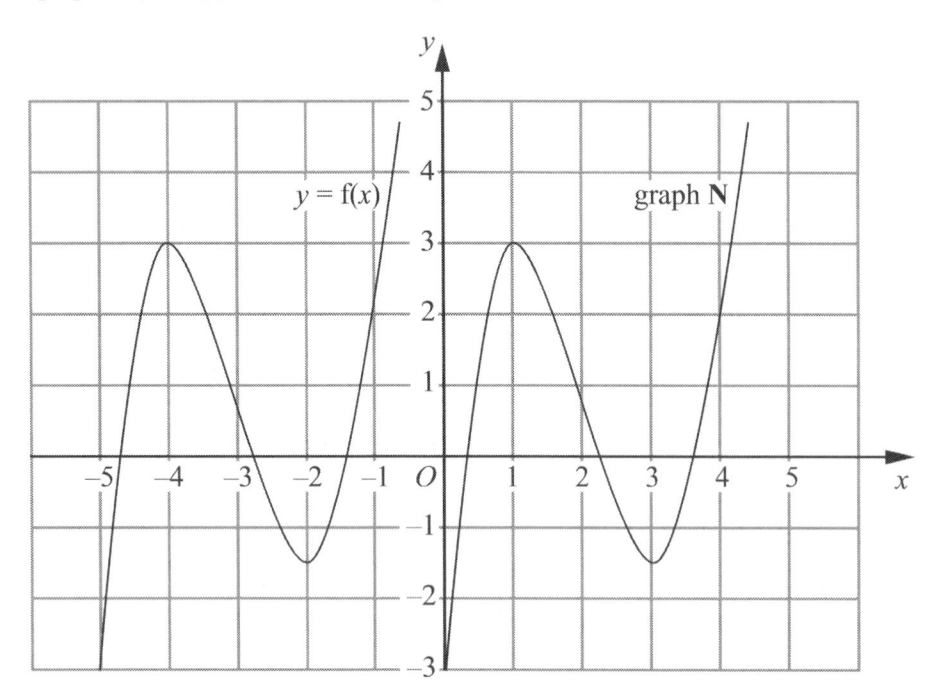

The graph N is a translation of the graph of y = f(x).

(a) Write down in terms of f, the equation of graph N

y = ...
(1)

The graph of y = f(x) has a maximum point at (−4, 3).

(b) Write down the coordinates of the maximum point of the graph of y = f(−x).

(...................... , ...)
(2)

(Total for Question 6 is 3 marks)

7 The graph of y = f(x) is shown on each of the grids.

   (a) On this grid, sketch the graph of y = f(x − 2)

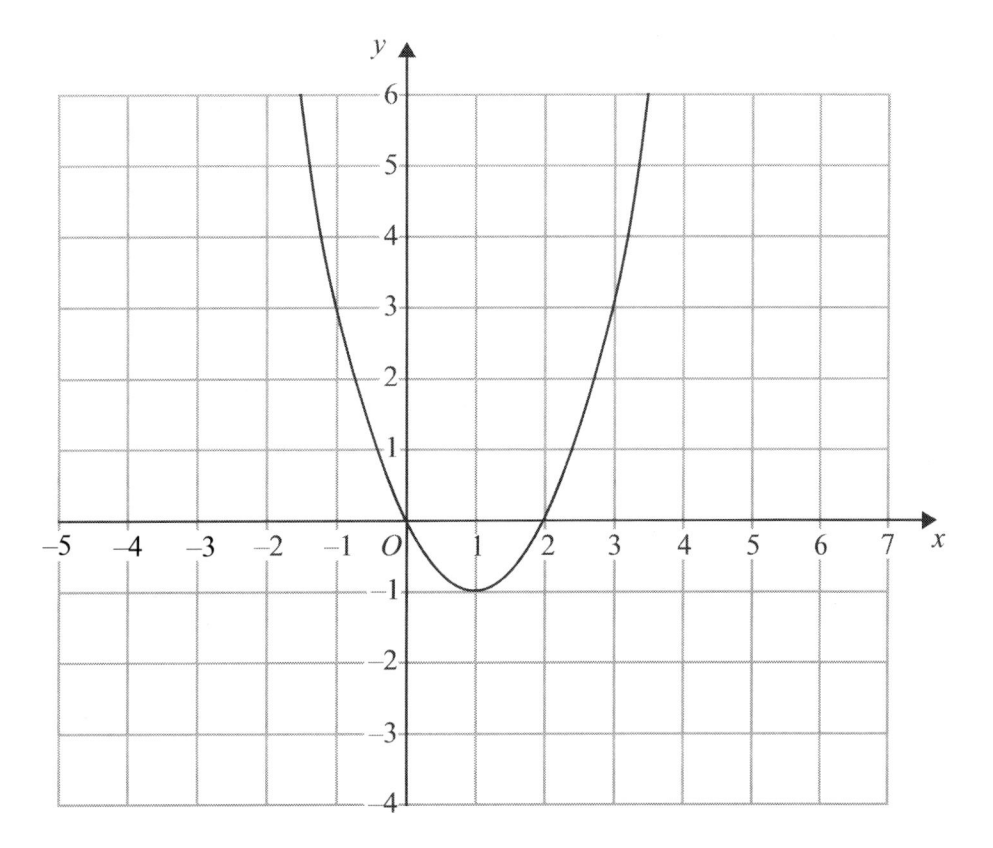

(2)

(b) On this grid, sketch the graph of $y = 2f(x)$

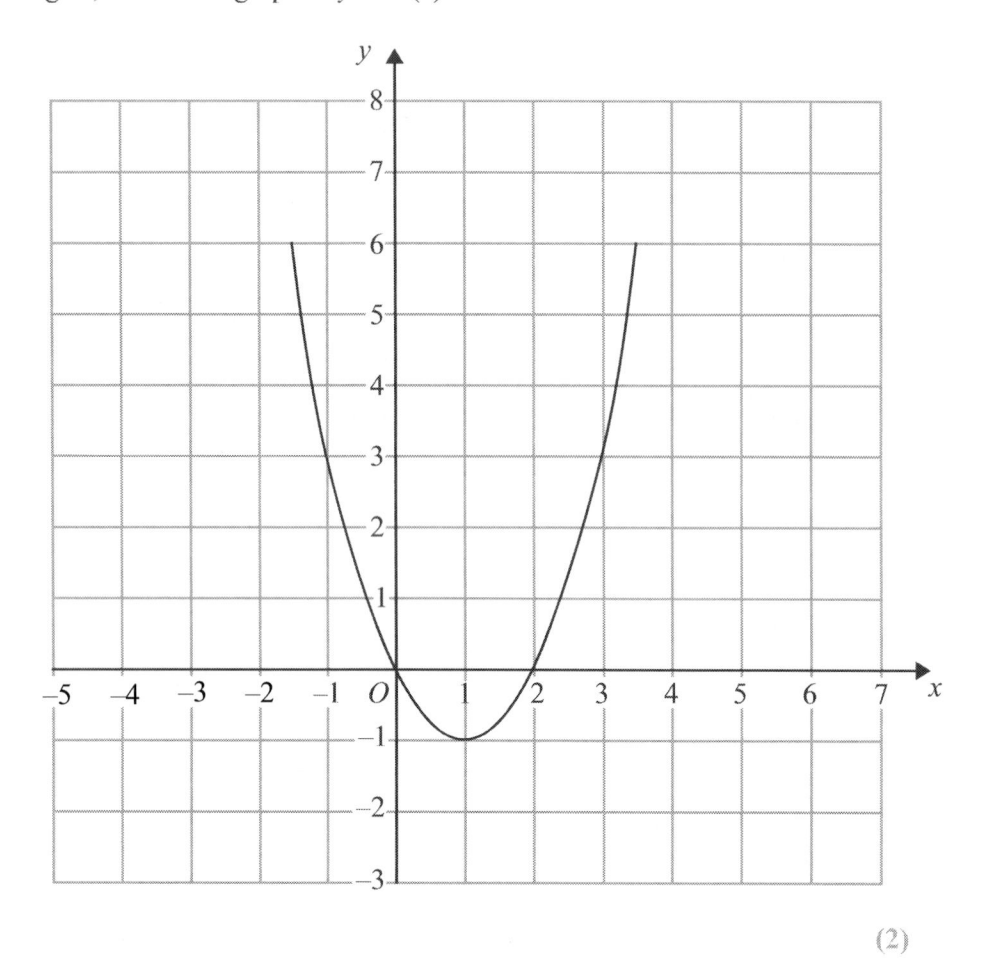

(2)

(Total for Question 7 is 4 marks)

8 The graph of $y = f(x)$ is drawn on the grid.

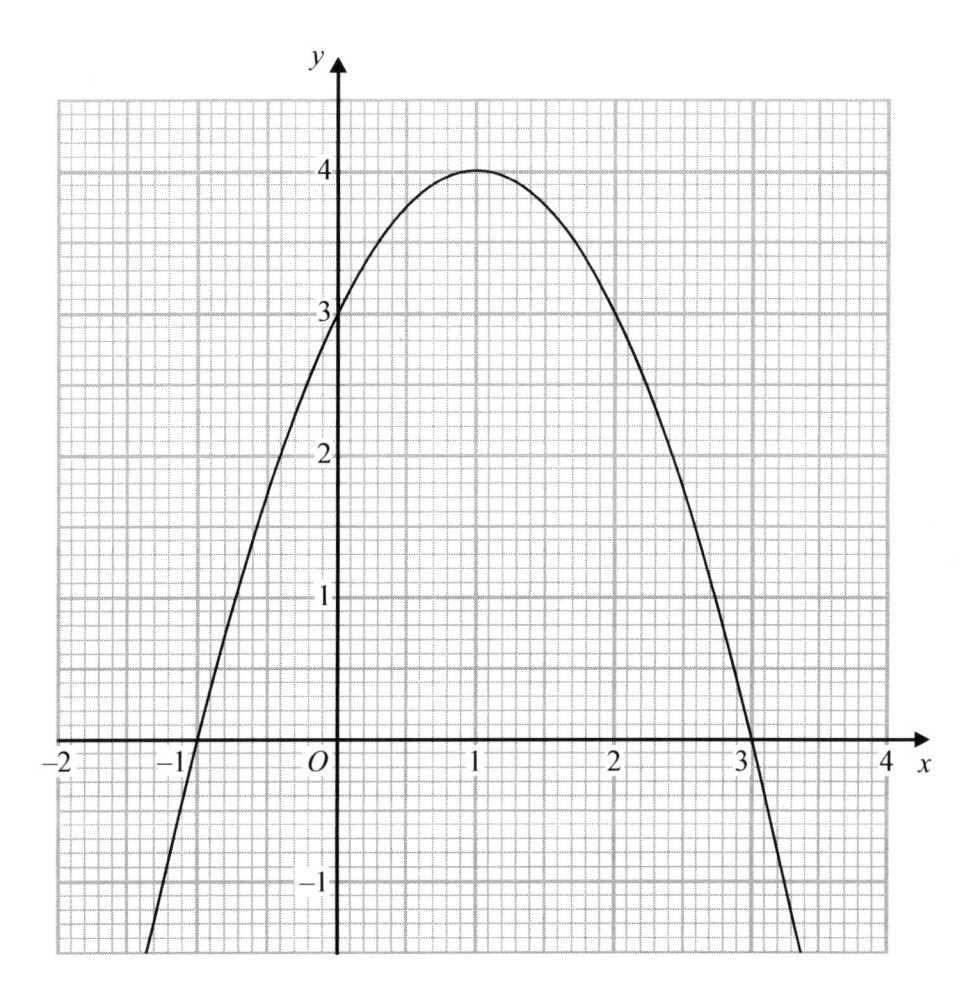

(a) Write down the coordinates of the turning point of the graph.

(............, ...)
(1)

(b) Write down the roots of $f(x) = 2$

....
(1)

(c) Write down the value of $f(0.5)$

....
(1)

(Total for Question 8 is 3 marks)

9 The graph of $y = f(x)$ is shown on both grids below.

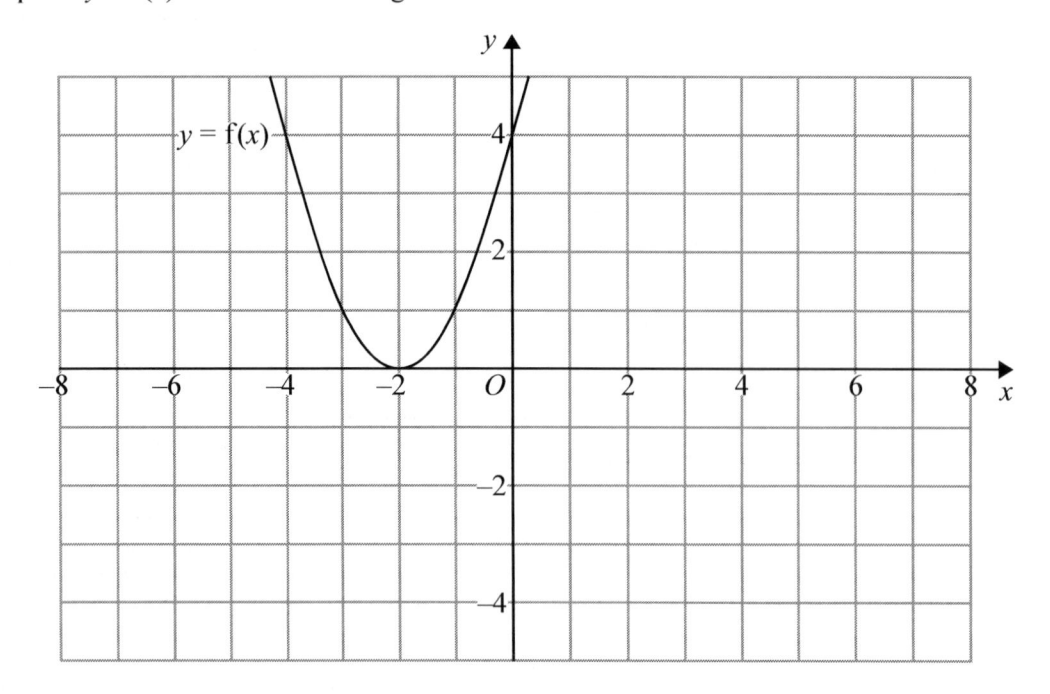

(a) On the grid above, sketch the graph of $y = f(-x)$

(1)

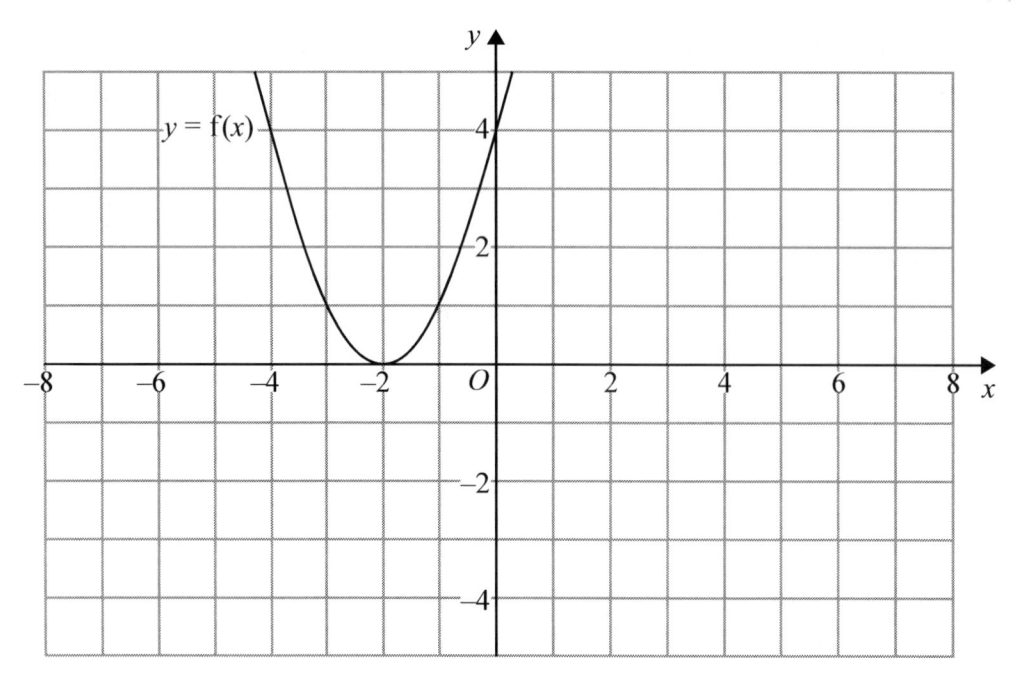

(b) On this grid, sketch the graph of $y = -f(x) + 3$

(1)

(Total for Question 9 is 2 marks)

Write your name here

Surname

Other names

In the style of:
**Pearson Edexcel
Level 1/Level 2 GCSE (9 - 1)**

Centre Number

Candidate Number

Mathematics
Transformations

Foundation Tier

GCSE style questions arranged by topic

Paper Reference
1MA1/1F

You must have: Ruler graduated in centimetres and millimetres, protractor, pair of compasses, pen, HB pencil, eraser.

Total Marks

Instructions

- Use **black** ink or ball-point pen.
- **Fill in the boxes** at the top of this page with your name, centre number and candidate number.
- Answer **all** questions.
- Answer the questions in the spaces provided
  – *there may be more space than you need.*
- **Calculators may not be used.**
- Diagrams are **NOT** accurately drawn, unless otherwise indicated.
- You must **show all your working out**.

Information

- The total mark for this paper is
- The marks for **each** question are shown in brackets
  – *use this as a guide as to how much time to spend on each question.*

Advice

- Read each question carefully before you start to answer it.
- Keep an eye on the time.
- Try to answer every question.
- Check your answers if you have time at the end.

Turn over ▶

© Peter Bland

1 Here are some patterns of circles.

(a) Shade **two** more circles to give this pattern symmetry in the mirror line.

(2)

(b) Shade **two** more circles to give this pattern symmetry in both mirror lines.

(2)

(c) Shade **four** more circles to give this pattern symmetry in the mirror line.

(2)

(Total for Question 1 is 6 marks)

2. The shape *ABCD* is drawn on a grid.

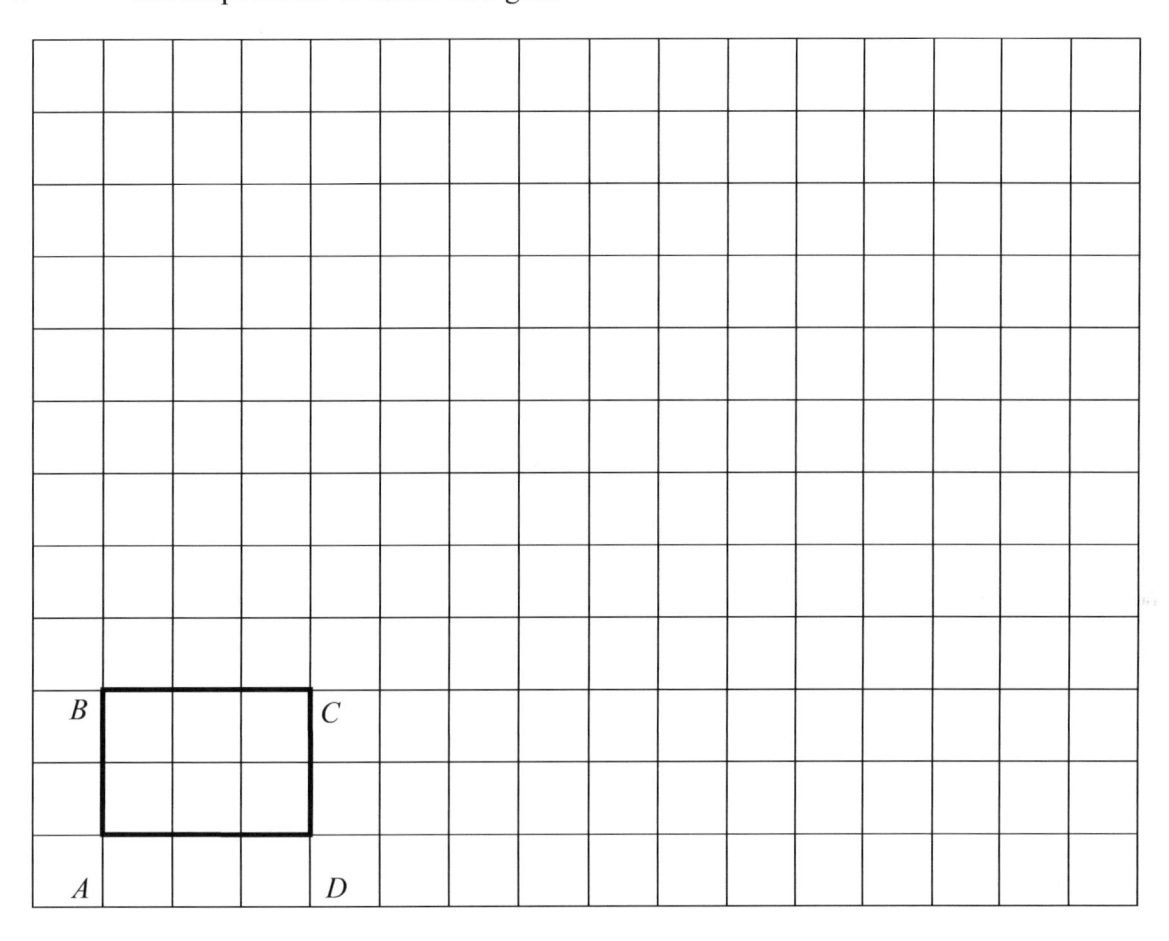

(a) Enlarge *ABCD* by scale factor 3.

(2)

(b) How many times bigger is the area of the enlarged shape than the area of *ABCD*?

....

(2)

(Total for Question 2 is 4 marks)

3

The number 71 is shaded on the grid.

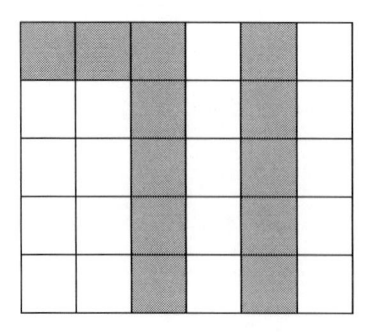

(a) What fraction of the grid is shaded?

Give your answer in its simplest form.

....

(3)

(b) The letter S is shaded on this grid.

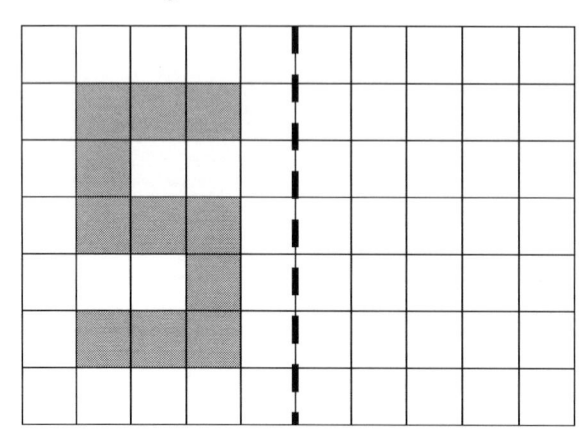

mirror line

Draw the reflection of the letter S in the mirror line.

(2)

3 (c) The number eight is drawn.

Write down the order of rotational symmetry.

....

(1)

(Total for Question 3 is 6 marks)

4

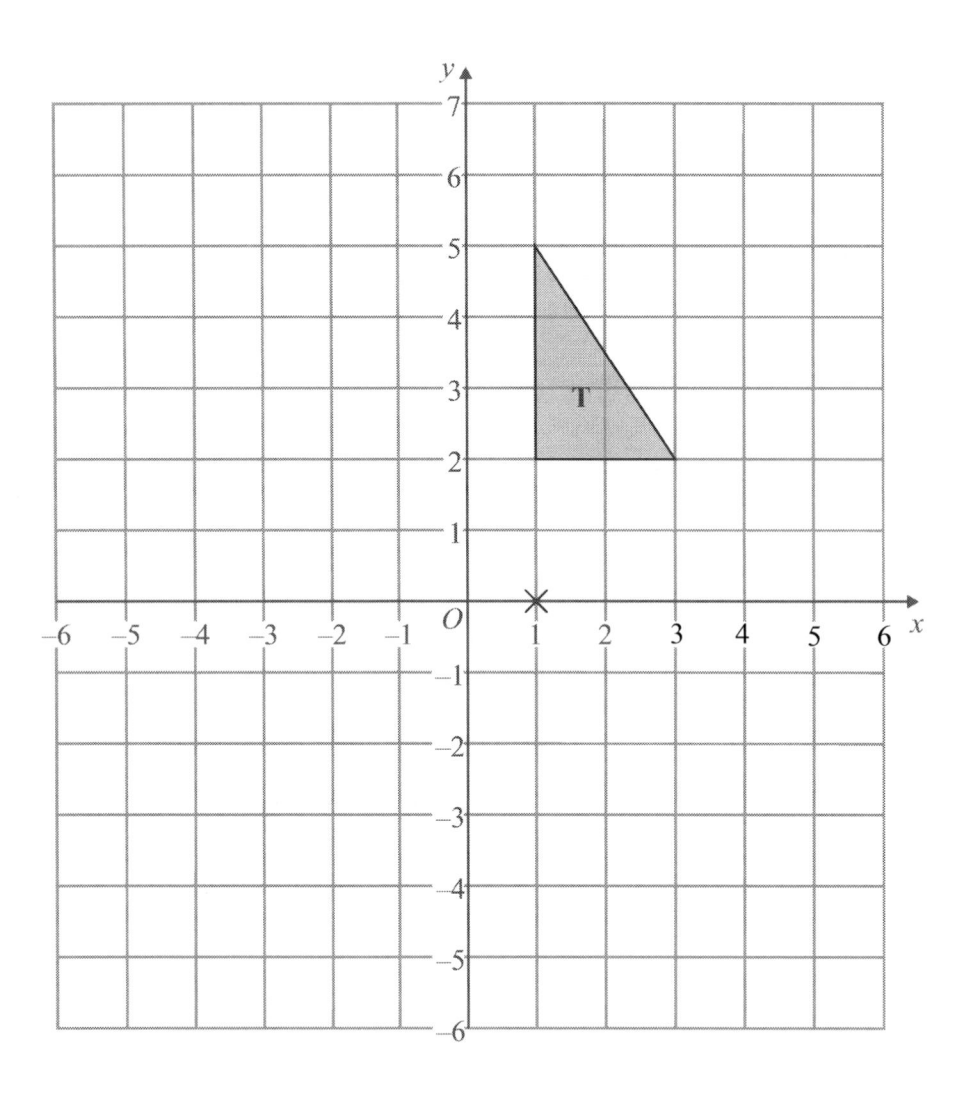

Triangle **T** has been drawn on the grid.

Rotate triangle **T** 90° clockwise about the point (1, 0).

Label the new triangle **A**.

(Total for Question 4 is 2 marks)

5

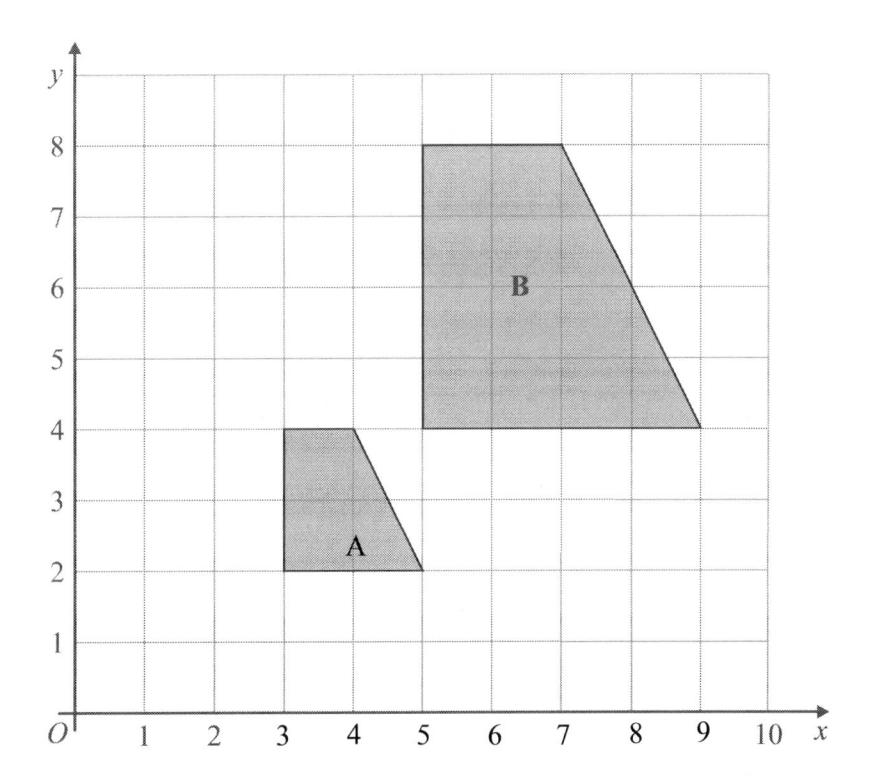

Describe fully the single transformation which maps shape **A** onto shape **B**.

....

....

(Total for Question 5 is 3 marks)

6

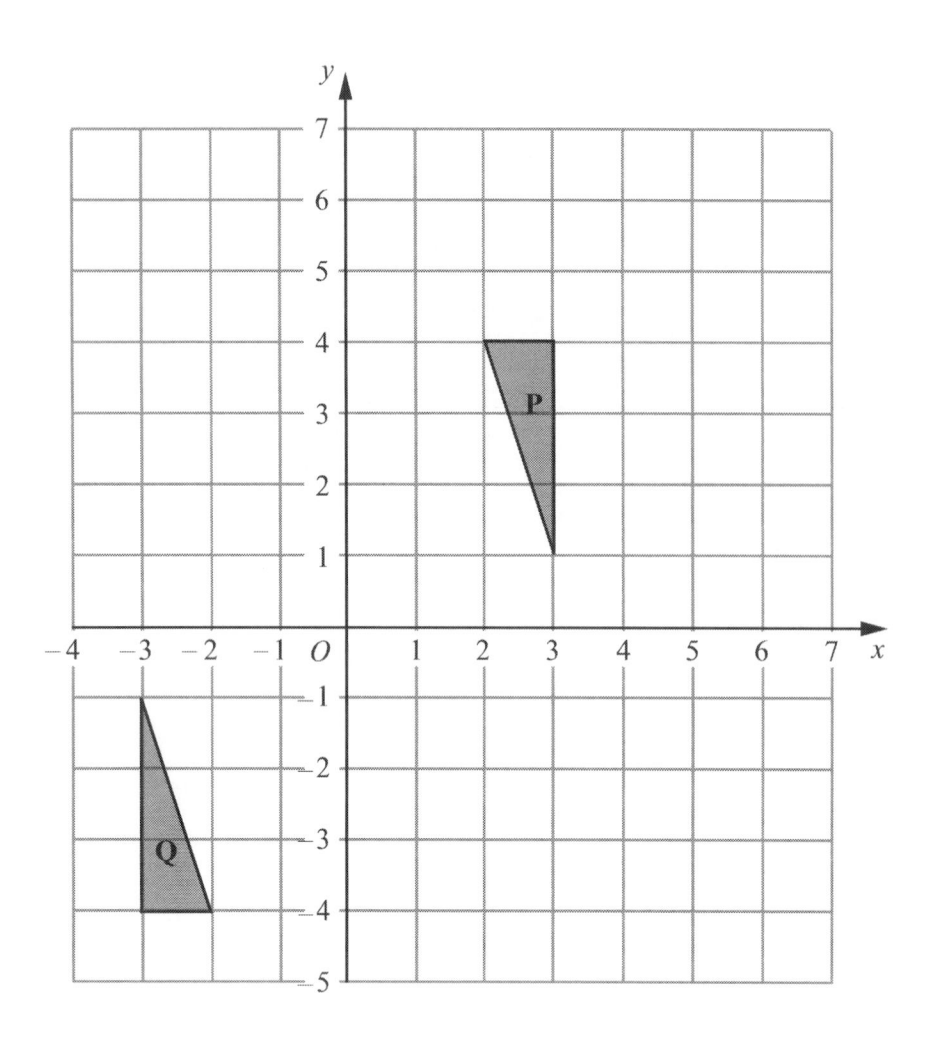

Triangle **P** and triangle **Q** are drawn on the grid.

(a) Describe fully the single transformation which maps triangle **P** onto triangle **Q**.

....

....

(3)

(b) Translate triangle **P** by the vector $\begin{pmatrix} 3 \\ 0 \end{pmatrix}$.

Label the new triangle **R**.

(1)

(Total for Question 6 is 4 marks)

7

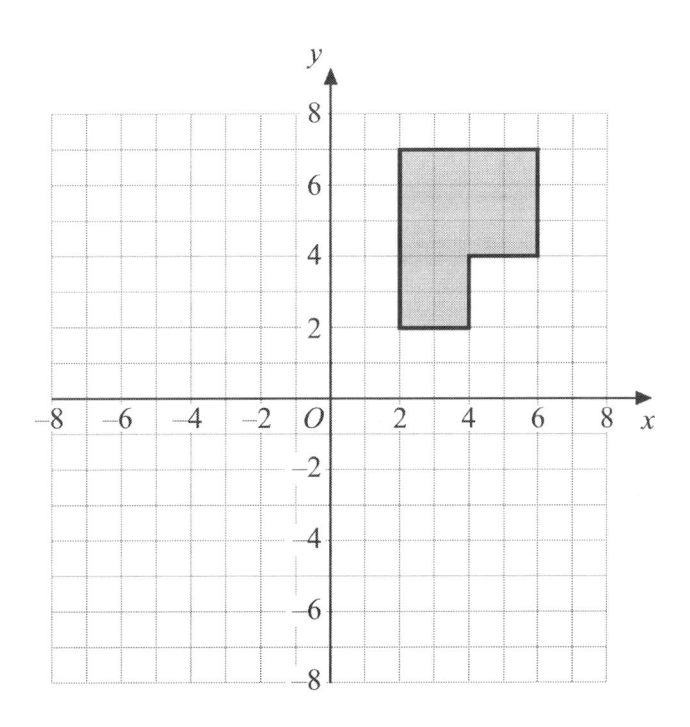

(a) Rotate the shaded shape 180° clockwise about the point O.

(2)

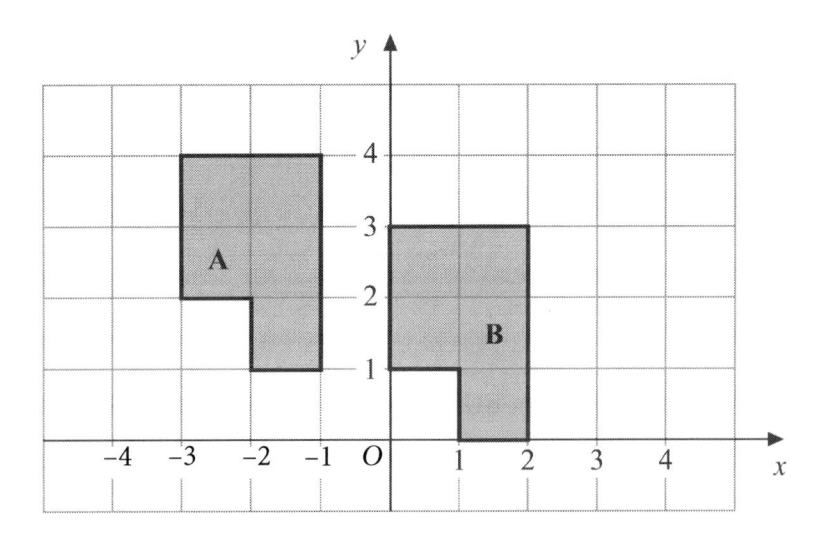

(b) Describe fully the single transformation that will map shape **A** onto shape **B**.

....

(2)

(Total for Question 7 is 4 marks)

8

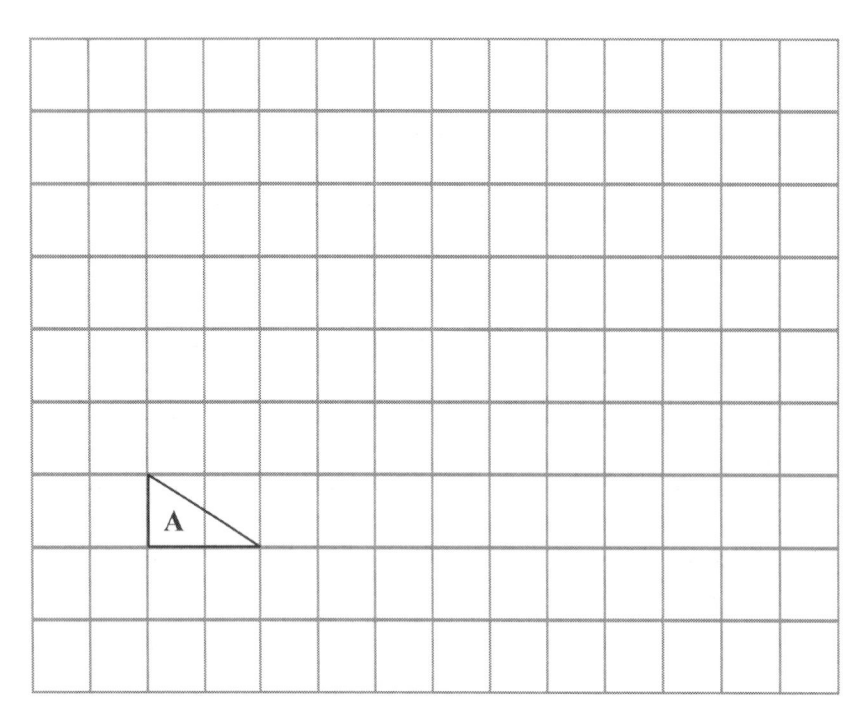

Triangle **A** has been drawn on a grid.

(a) On the grid, draw an enlargement of the triangle **A** with a scale factor 3.

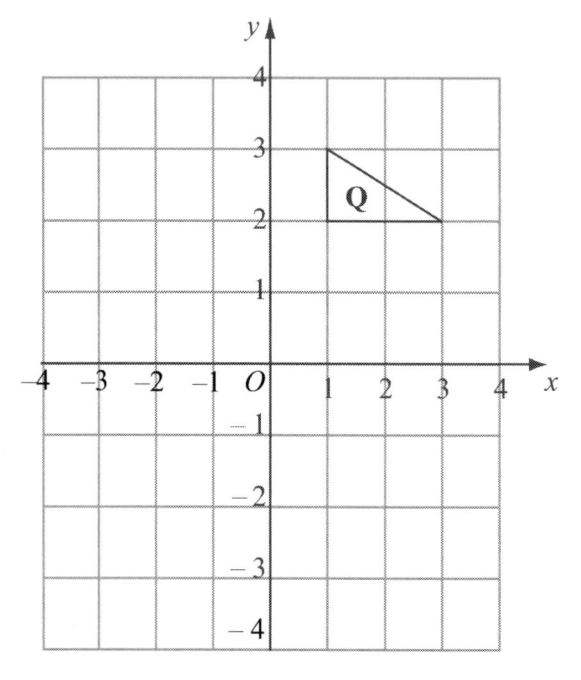

Triangle Q has been drawn on a grid.
(b) On the grid, rotate triangle Q 90° clockwise, centre O.

9

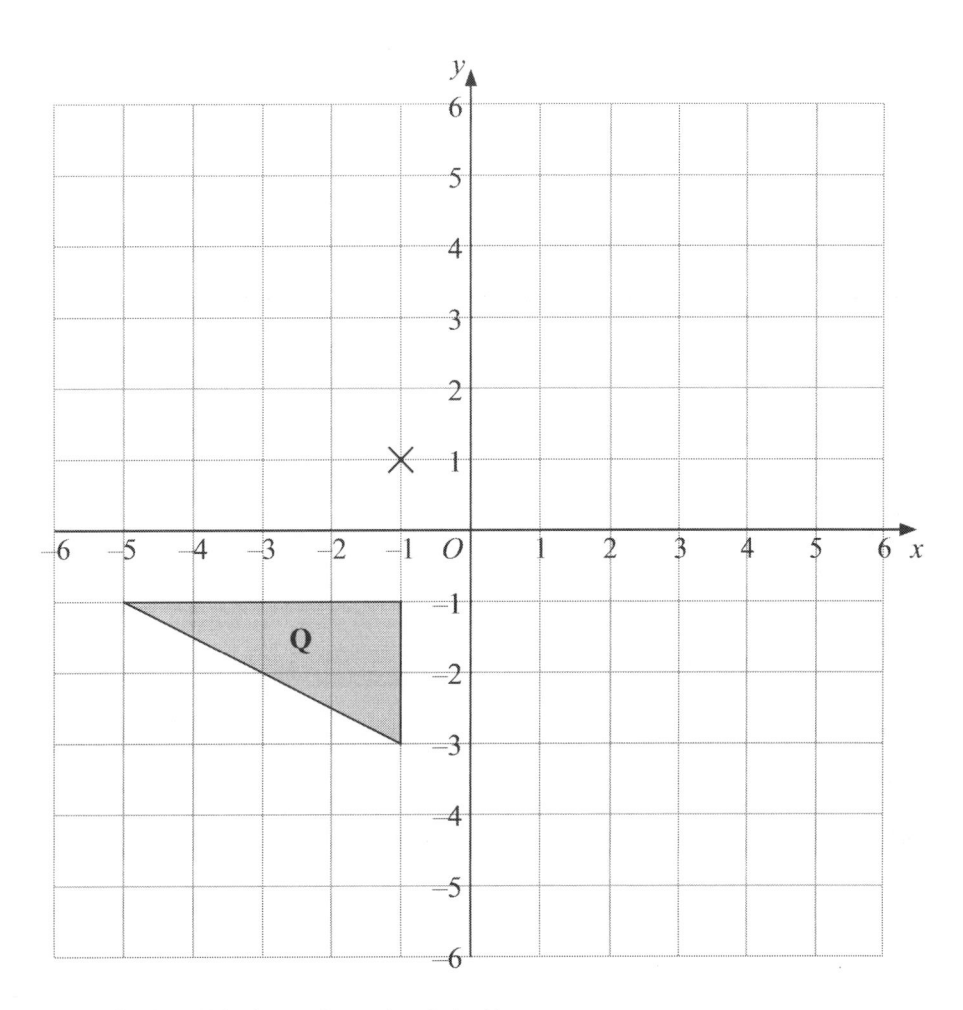

(a) Rotate triangle **Q** 180° about the point (–1, 1).

Label the new triangle **A**.

(2)

(b) Translate triangle **Q** by the vector $\begin{pmatrix} 6 \\ -1 \end{pmatrix}$.

Label the new triangle **B**.

(1)

(Total for Question 9 is 3 marks)

10

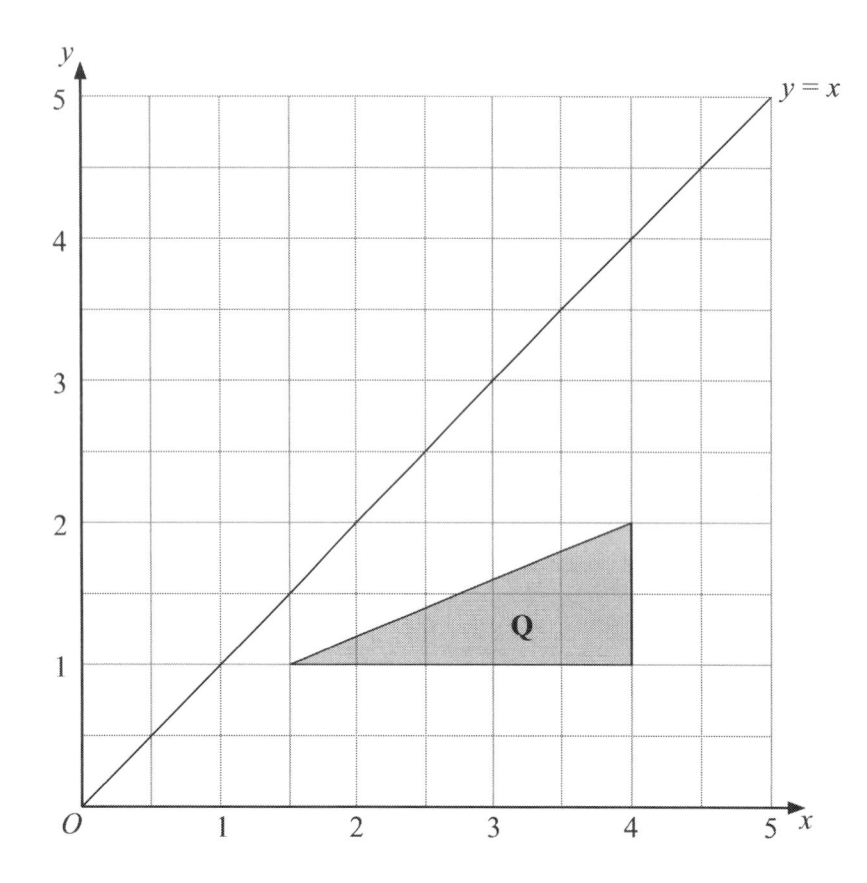

(c) Reflect triangle **Q** in the line $y = x$.

Label the new triangle **C**.

(5)

(Total for Question 10 is 5 marks)

11

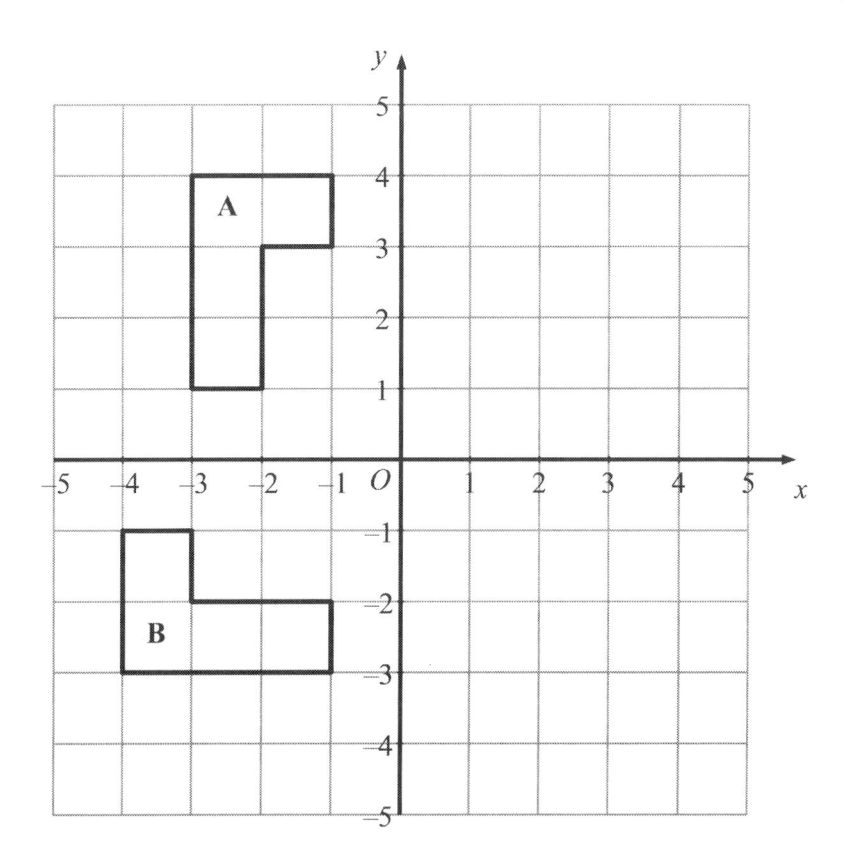

(a) Reflect shape **A** in the *y* axis.

(2)

(b) Describe fully the **single** transformation which takes shape **A** to shape **B**.

....

(3)

(Total for Question 11 is 5 marks)

12

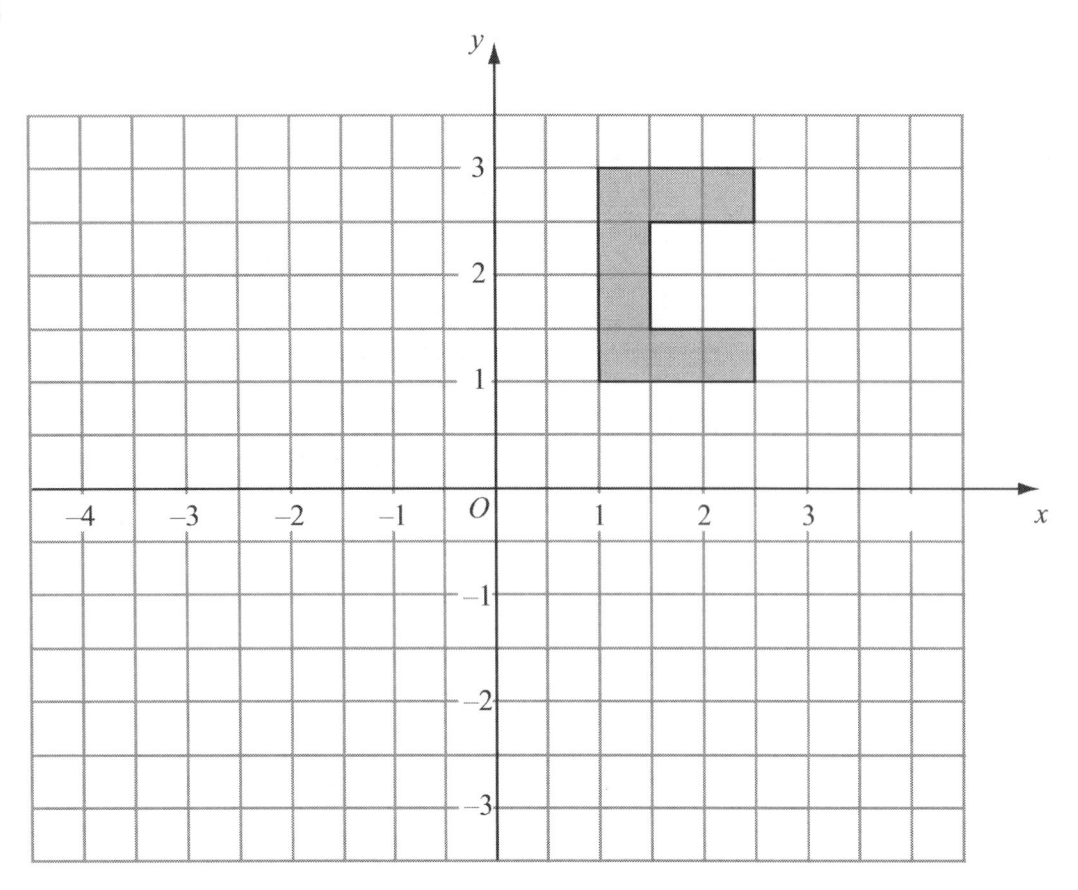

Rotate the shape 180° centre O.

(Total for Question 12 is 2 marks)

13

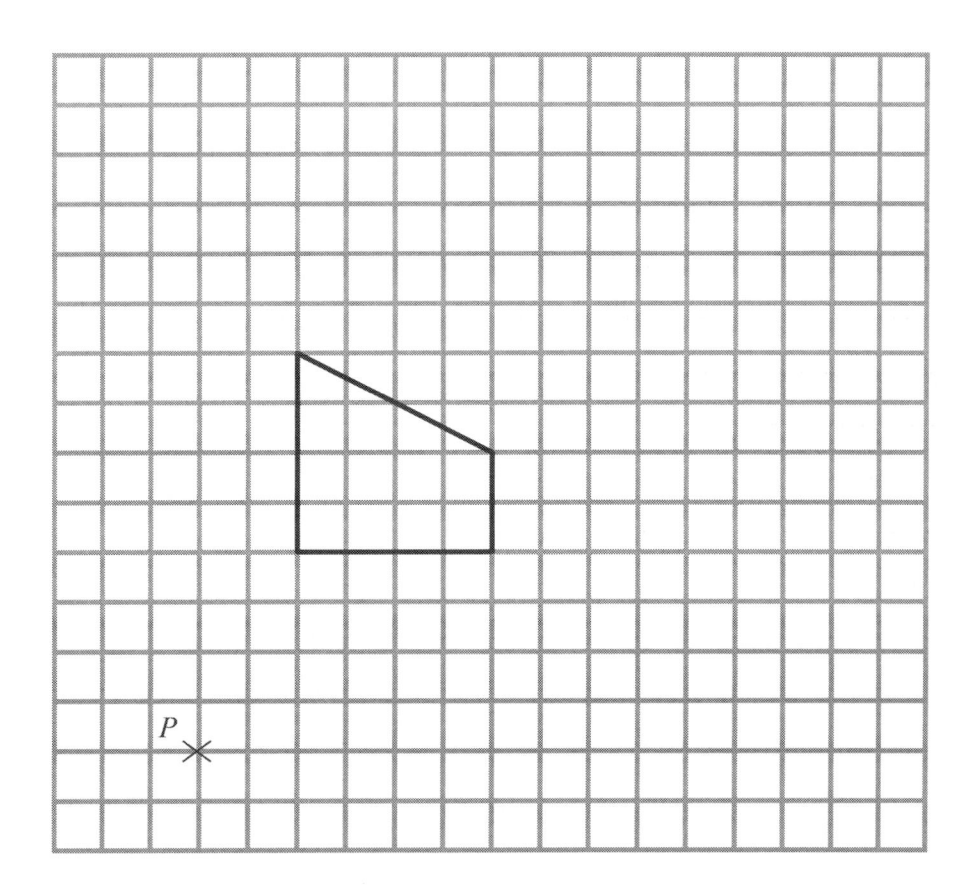

On the grid, enlarge the shape with a scale factor of $\frac{1}{2}$, centre P.

(Total for Question 13 is 3 marks)

14

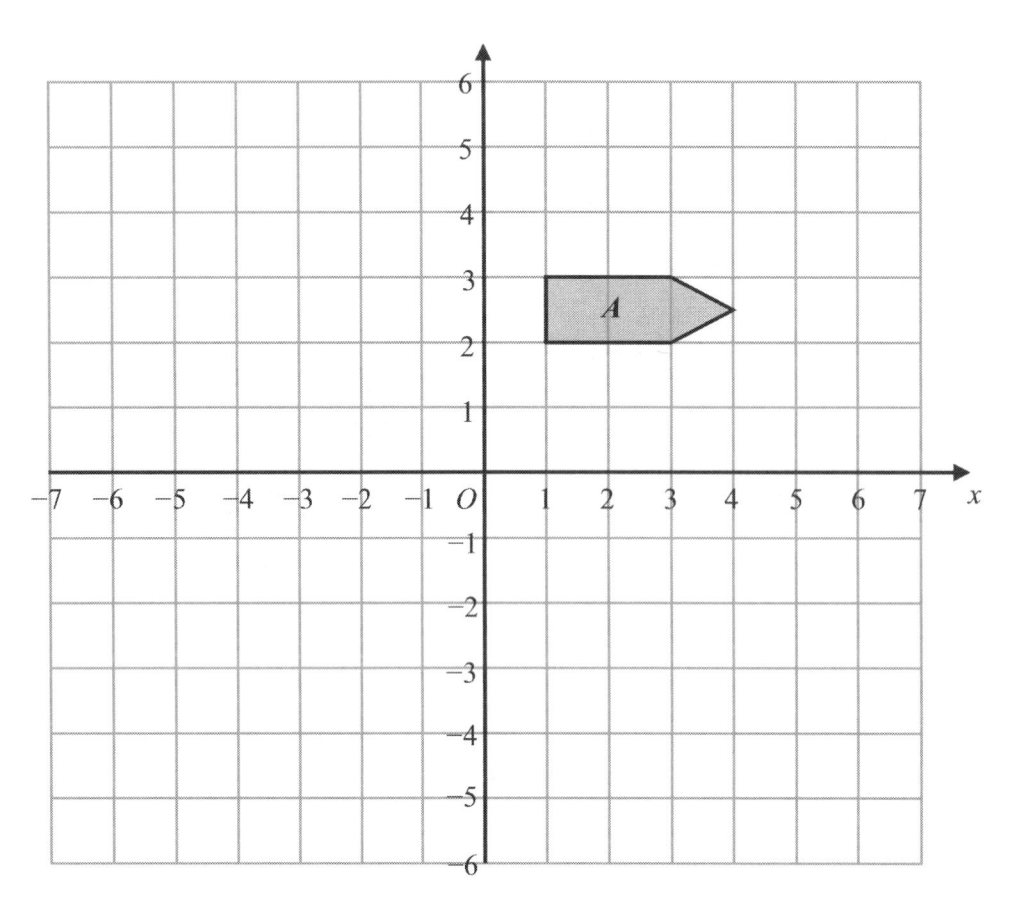

(a) On the grid above, reflect shape *A* in the line $x = -2$

(2)

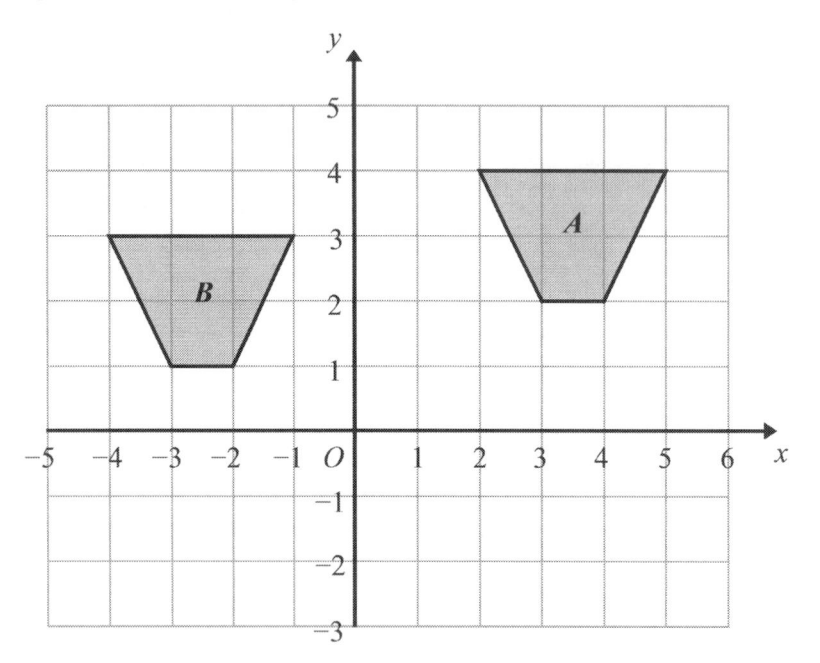

(b) Describe fully the single transformation that will map shape *A* onto shape *B*.

....

....

(2)

(Total for Question 14 is 4 marks)

15

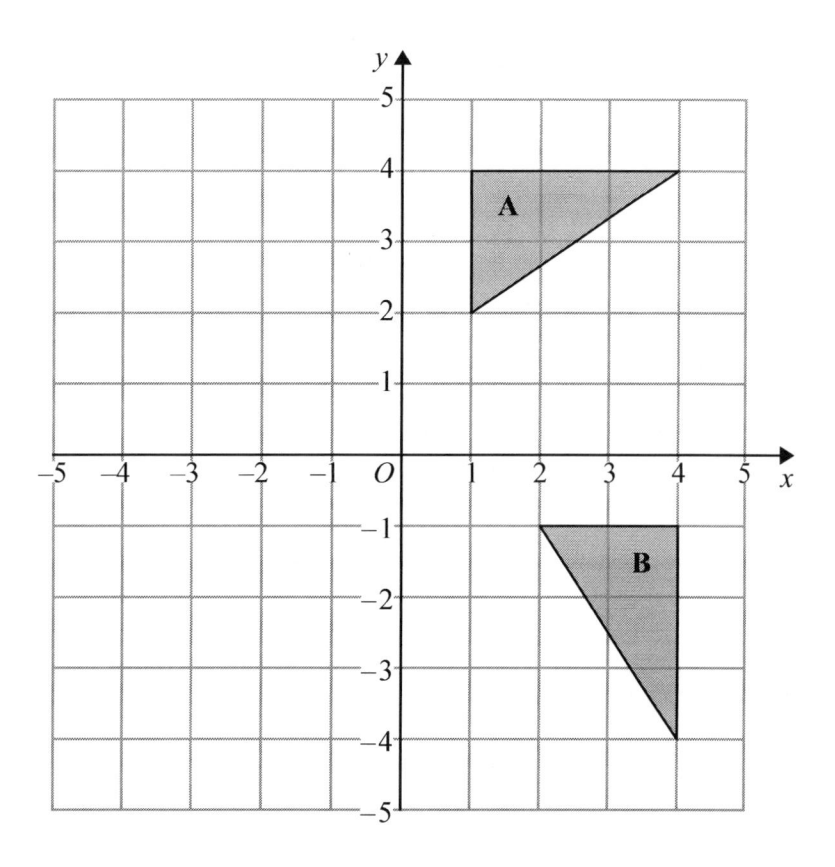

Describe fully the single transformation that maps triangle **A** onto triangle **B**.

....

....

(Total for Question 15 is 2 marks)

16

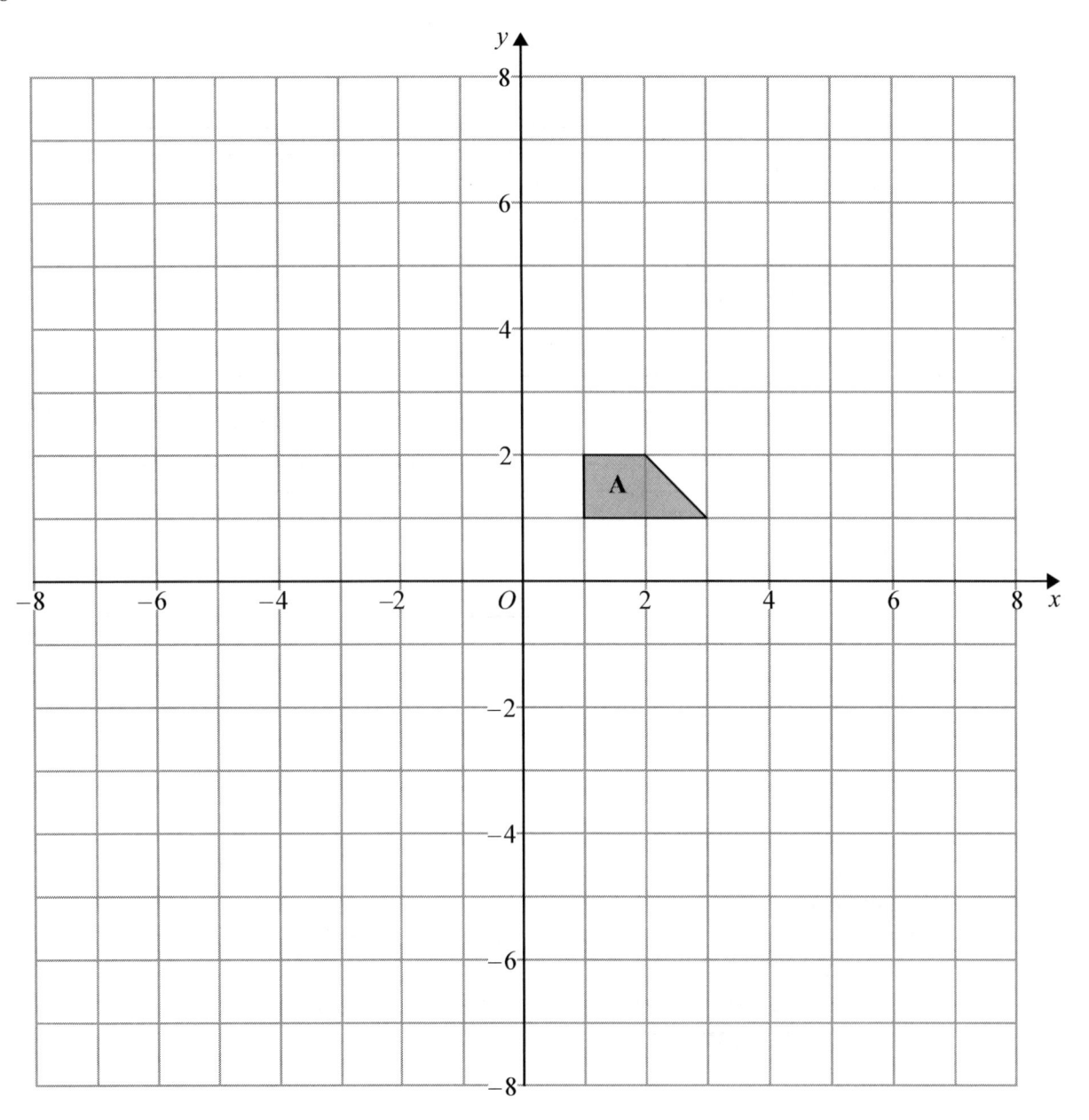

(a) Enlarge shape **A** by scale factor −2, centre (0, 0)
Label your image **B**.

(2)

(b) Describe fully the single transformation that will map shape **B** onto shape **A**.

....

(1)

(Total for Question 16 is 3 marks)

Write your name here

Surname

Other names

In the style of:
**Pearson Edexcel
Level 1/Level 2 GCSE (9 - 1)**

Centre Number

Candidate Number

Mathematics
Trigonometry

Higher Tier

GCSE style questions arranged by topic

Paper Reference
1MA1/2H

You must have: Ruler graduated in centimetres and millimetres, protractor, pair of compasses, pen, HB pencil, eraser, calculator.

Total Marks

Instructions

- Use **black** ink or ball-point pen.
- **Fill in the boxes** at the top of this page with your name, centre number and candidate number.
- Answer **all** questions.
- Answer the questions in the spaces provided
  – *there may be more space than you need.*
- **Calculators may be used.**
- If your calculator does not have a π button, take the value of π to be 3.142 unless the question instructs otherwise.
- Diagrams are **NOT** accurately drawn, unless otherwise indicated.
- You must **show all your working out**.

Information

- The total mark for this paper is
- The marks for **each** question are shown in brackets
  – *use this as a guide as to how much time to spend on each question.*

Advice

- Read each question carefully before you start to answer it.
- Keep an eye on the time.
- Try to answer every question.
- Check your answers if you have time at the end.

Turn over ▶

© Peter Bland

1. *ABC* is a right-angled triangle. *AB* = 18 cm and *BC* = 6 cm.
The line *BD* bisects the angle *ABC*.

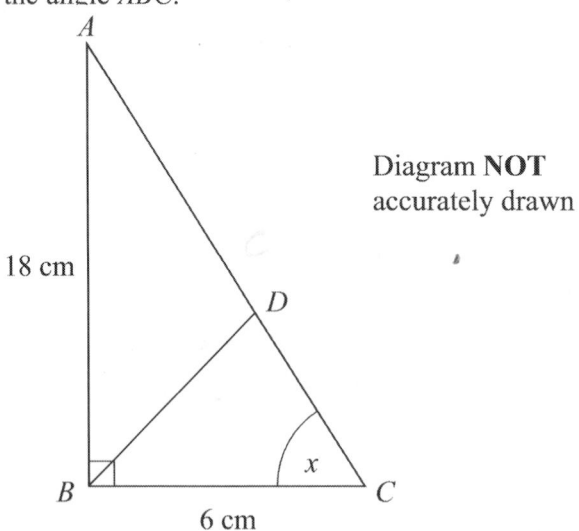

Diagram **NOT** accurately drawn

(a) Write down the value of tan *x*.

.... (1)

(b) Calculate the length *BD*.

$AC =$

$\dfrac{a}{\sin A} = \dfrac{b}{\sin B}$

$= \dfrac{6}{\sin 45} = \dfrac{18.9...}{x}$

$\dfrac{6}{\sin 45} \times 18.9... = x$

$x = 16$

.... cm (5)

(Total for Question 1 is 6 marks)

2 Here is a right-angled triangle.

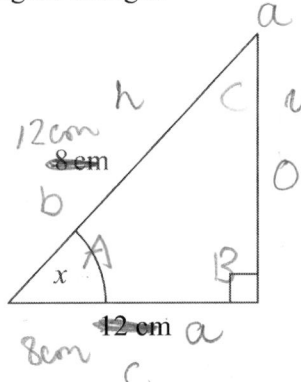

Diagram **NOT** accurately drawn

Student annotations on diagram: sides labeled a, b, c with 12 cm, 8 cm, and $4\sqrt{5}$; angle x at vertex A; right angle marked.

$b^2 = a^2 - c^2$
$b^2 = 12^2 - 8^2$
$b = \sqrt{80}$
$b = 4\sqrt{5}$

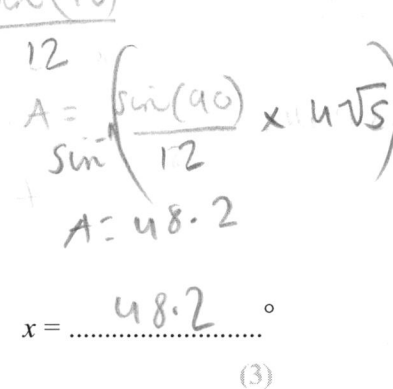

$\dfrac{\sin A}{4\sqrt{5}} = \dfrac{\sin(90)}{12}$

$A = \sin^{-1}\left(\dfrac{\sin(90)}{12} \times 4\sqrt{5}\right)$

$A = 48.2$

(a) Calculate the size of the angle marked x.
Give your answer correct to 1 decimal place.

$x = $ ......48.2...... °

(3)

Here is another right-angled triangle.

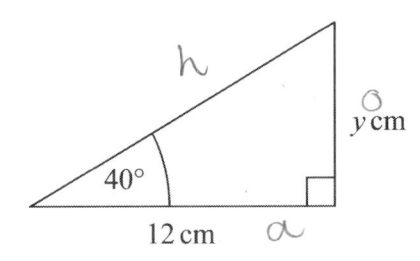

Diagram **NOT** accurately drawn

Triangle with 40° angle, base 12 cm, opposite side y cm, hypotenuse h.

(b) Calculate the value of y.
Give your answer correct to 1 decimal place.

$\tan(40) = \dfrac{y}{12}$

$y = \tan(40) \times 12$

$y = 10.06919557$

$y = 10.1$

$y = $ ......10.1......

(3)

(Total for Question 2 is 6 marks)

379

3

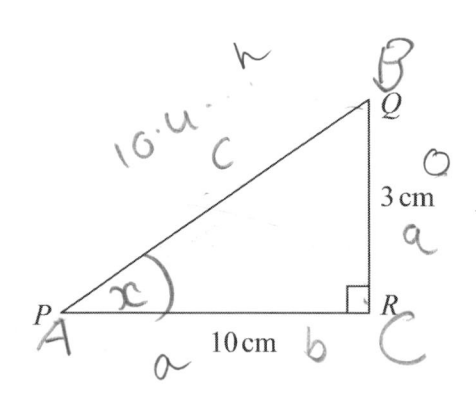

Diagram **NOT** accurately drawn

PQR is a right-angled triangle.

QR = 3 cm
PR = 10 cm

Work out the size of angle *RPQ*.
Give your answer correct to 3 significant figures.

$a^2 + b^2 = c^2$
$3^2 + 10^2 = c^2$
$9 + 100 = c^2$
$c^2 = 109$
$c = \sqrt{109}$
$c = 10.4403065 1$
$(c = 10.4)$

SOH CAH TOA

$\tan(x) = \frac{3}{10}$
$x = \tan^{-1}\left(\frac{3}{10}\right)$
$= 16.70$

$t = \frac{o}{a}$
$t = \frac{3}{10}$
$t = 0.3$

.......16.7....... °

(Total for Question 3 is 3 marks)

4

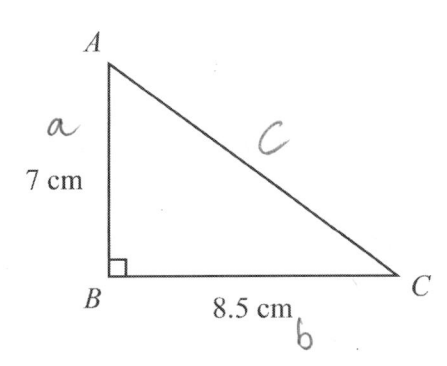

Diagram NOT accurately drawn

ABC is a right-angled triangle.
AB = 7 cm,
BC = 8.5 cm.

(a) Work out the area of the triangle.

$\frac{1}{2} ab \sin C$

$\frac{1}{2} \times 7 \times 8.5 \times \sin(90)$

$= 29.75$

..........29.75.......... cm²
(2)

(b) Work out the length of *AC*.
Give your answer correct to 2 decimal places.

$a^2 + b^2 = c^2$
$= 7^2 + 8.5^2$
$= 49 + 72.25$
$= \sqrt{121.25}$
$= 11.01135777$

..........11.0.......... cm
(3)

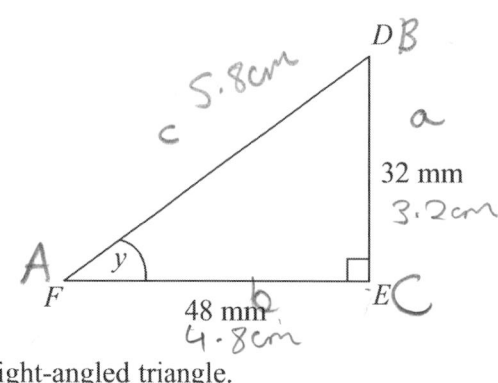

Diagram **NOT** accurately drawn

Handwritten annotations on diagram:
- c 5.8cm
- D B
- a
- 32 mm / 3.2cm
- A / y
- F / 48 mm / 4.8cm / b / E C

DEF is another right-angled triangle.
DE = 32 mm,
FE = 48 mm.

(c) Calculate the size of angle *y*.
Give your answer correct to 1 decimal place.

Handwritten working (crossed out):

$$\frac{\sin A}{a} = \frac{\sin C}{c}$$

$$\frac{\sin(x)}{3.2} = \frac{\sin(90)}{45.8}$$

$$\sin(x) = \frac{\sin(90)}{45.8} \times 3.2$$

Handwritten working (right side):

$a^2 + b^2 = c^2$

$3.2^2 + 4.8^2 = c^2$

$c^2 = 10.24 + 23.04$

$c^2 = \sqrt{33.28}$

$c = 5.768882041$

$c = 5.8 cm (2dp)$

.......................°

(3)

(Total for Question 4 is 8 marks)

Handwritten at bottom: SOH CAH TOA

382

5

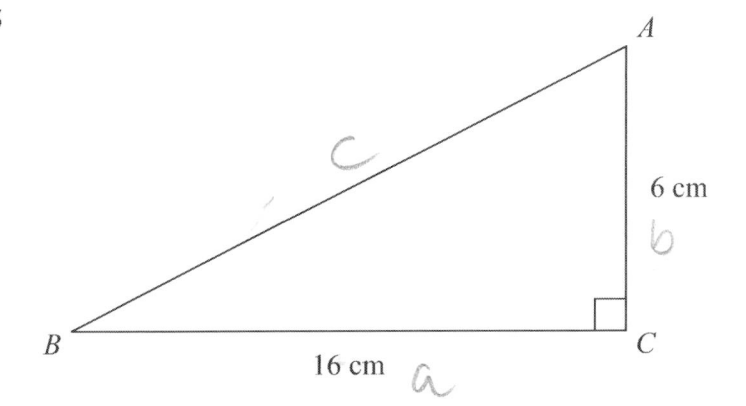

Diagram **NOT** accurately drawn

ABC is a right-angled triangle.
AC = 6 cm.
BC = 16 cm.

(a) Work out the area of triangle *ABC*.

$\frac{1}{2}ab\sin C$

$\frac{1}{2} \times 16 \times 6 \times \sin(90)$
$= 48$

.... 48 .... cm²
(2)

(b) Calculate the length of *AB*.
Give your answer correct to 2 decimal places.

$a^2 + b^2 = c^2$
$16^2 + 6^2 = c^2$
$c^2 = 292$
$c = \sqrt{292}$
$c = 17.1$

.... 17.1 .... cm
(3)

(Total for Question 5 is 5 marks)

6

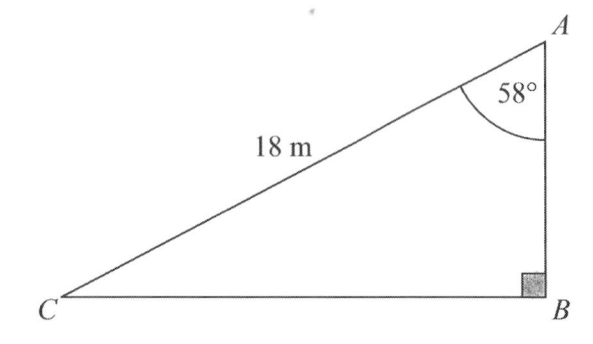

Diagram **NOT** accurately drawn

ABC is a right-angled triangle.
AC = 18 m.
Angle *CAB* = 58°

Calculate the length of *AB*.
Give your answer correct to 3 significant figures.

.... m

(Total for Question 6 is 3 marks)

7.

ABC is a triangle.
AB = 8 cm
BC = 15 cm
Angle *ABC* = 112°

Calculate the area of the triangle.
Give your answer correct to 3 significant figures.

.... cm^2

(Total for Question 7 is 3 marks)

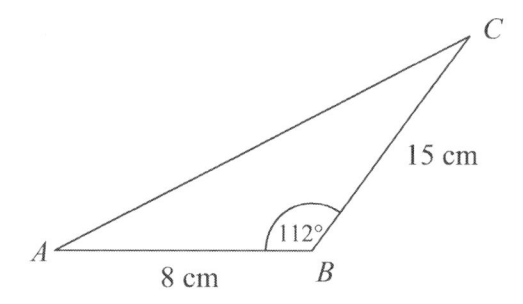

8 Town *B* is 4.6 km due West of town *C*.
Town *A* is 2.3 km due North of town *B*.

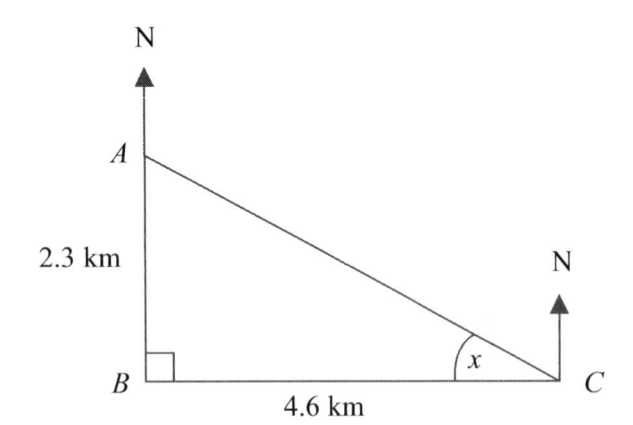

Diagram **NOT** accurately drawn

(a) Calculate the size of the angle marked *x*.
Give your answer correct to 3 significant figures.

x = ... °

(3)

(b) Find the bearing of town *C* from town *A*.
Give your answer correct to 3 significant figures.

.... °

(1)

(Total for Question 8 is 4 marks)

9

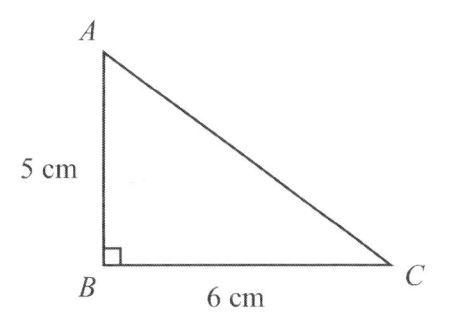

*Diagram **NOT** accurately drawn*

ABC is a right-angled triangle.
AB = 5 cm,
BC = 6 cm.

(a) Work out the area of the triangle.

.... cm^2
(2)

(b) Work out the length of *AC*.
Give your answer correct to 2 decimal places.

.... cm
(3)

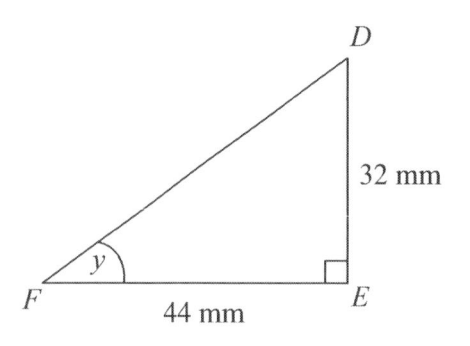

Diagram **NOT** accurately drawn

DEF is another right-angled triangle.
DE = 32 mm,
FE = 44 mm.

(c) Calculate the size of angle *y*.
   Give your answer correct to 1 decimal place.

.... °

(3)

(Total for Question 9 is 8 marks)

10

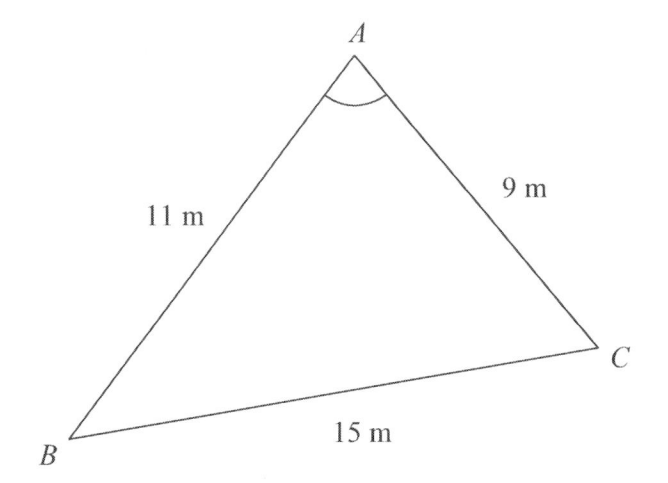

Diagram **NOT** accurately drawn

ABC is a triangle.
AB = 11 m.
AC = 9 m.
BC = 15 m.

Calculate the size of angle *BAC*.
Give your answer correct to one decimal place.

.... °

(Total for Question 10 is 3 marks)

11

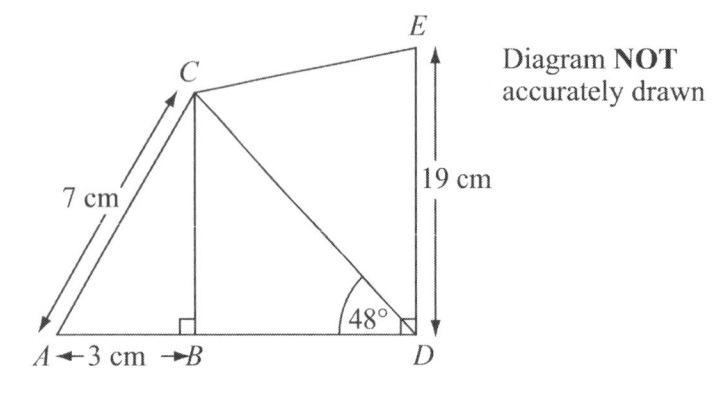

Diagram **NOT** accurately drawn

$AC = 7$ cm. $AB = 3$ cm. $DE = 19$ cm.

Angle ABC = angle CBD = angle BDE = 90°
Angle BDE = 48°
Calculate the length of CD

Give your answer correct to 3 significant figures.

.... cm

(4)

(b) Calculate the length of CE.
   Give your answer correct to 3 significant figures.

.... cm

(3)

(Total for Question 11 is 7 marks)

12

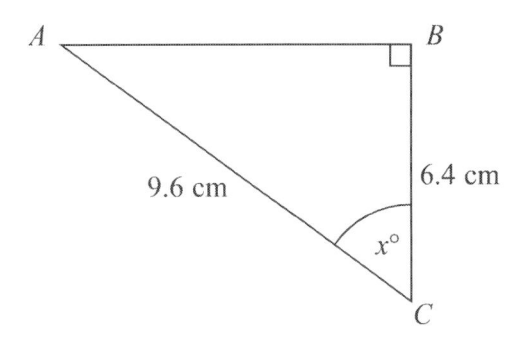

Diagram **NOT** accurately drawn

ABC is a right-angled triangle.

AC = 9.6 cm.
BC = 6.4 cm.

Calculate the size of the angle marked $x°$.
Give your answer correct to 1 decimal place.

.... °

(Total for Question 12 is 3 marks)

13 Triangles *ABD* and *BCD* are right-angled triangles.

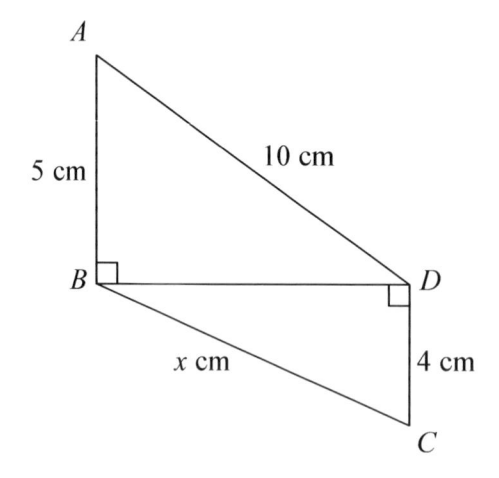

5 cm, 10 cm, x cm, 4 cm

Work out the value of x.
Give your answer correct to 2 decimal places.

....

(Total for Question 13 is 4 marks)

14 The diagram shows a design for a zipwire.

The zipwire will run between the top of two vertical posts, *AB* and *CD*.

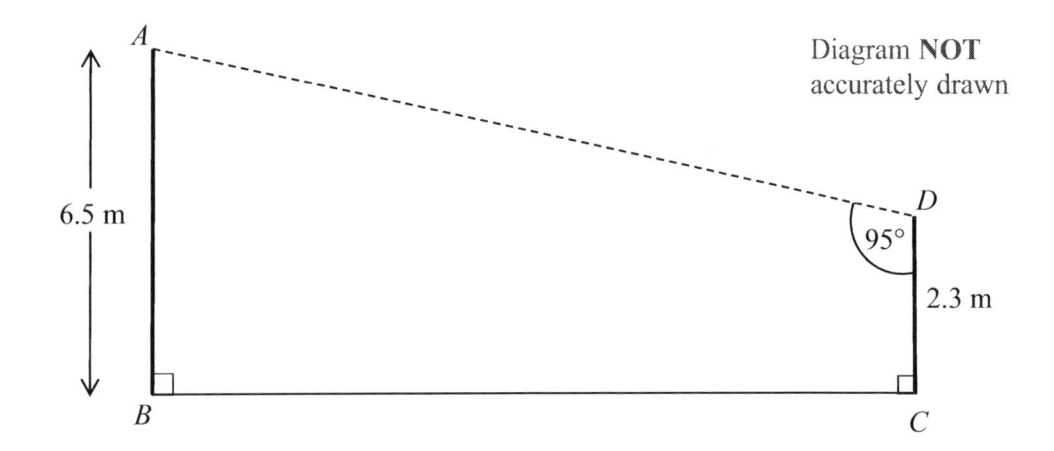

Diagram **NOT** accurately drawn

- *AB* = 6.5 m
- *CD* = 2.3 m
- Angle at *D* = 95°

Work out the distance *AD*.

.... m

(Total for Question 14 is 4 marks)

15 In the triangle, angle *y* is obtuse.

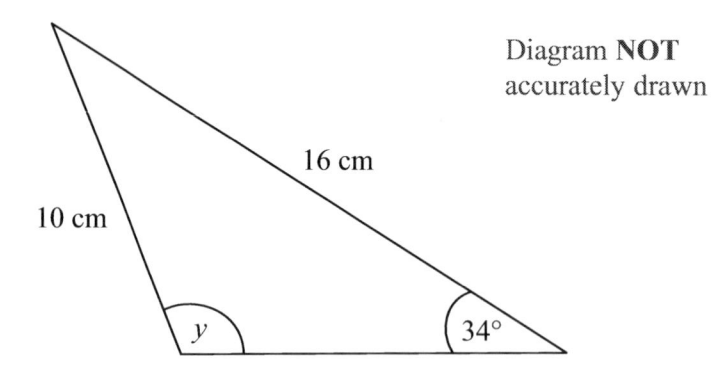

Diagram **NOT** accurately drawn

16 cm

10 cm

y

34°

Work out the size of angle *y*.

.................................°

(Total for Question 15 is 3 marks)

Write your name here

Surname

Other names

In the style of:
Pearson Edexcel
Level 1/Level 2 GCSE (9 - 1)

Centre Number

Candidate Number

Mathematics

Vectors

Higher Tier

GCSE style questions arranged by topic

Paper Reference
1MA1/3H

You must have: Ruler graduated in centimetres and millimetres, protractor, pair of compasses, pen, HB pencil, eraser, calculator.

Total Marks

Instructions

- Use **black** ink or ball-point pen.
- **Fill in the boxes** at the top of this page with your name, centre number and candidate number.
- Answer **all** questions.
- Answer the questions in the spaces provided
  – *there may be more space than you need.*
- **Calculators may be used.**
- If your calculator does not have a π button, take the value of π to be 3.142 unless the question instructs otherwise.
- Diagrams are **NOT** accurately drawn, unless otherwise indicated.
- You must **show all your working out**.

Information

- The total mark for this paper is
- The marks for **each** question are shown in brackets
  – *use this as a guide as to how much time to spend on each question.*

Advice

- Read each question carefully before you start to answer it.
- Keep an eye on the time.
- Try to answer every question.
- Check your answers if you have time at the end.

Turn over ▶

© Peter Bland

1

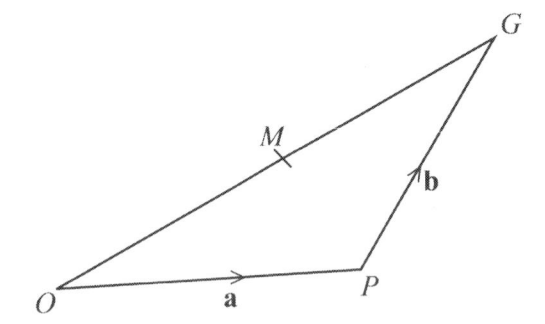

Diagram **NOT** accurately drawn

OGP is a triangle.

M is the midpoint of OG.

$\overrightarrow{OP} = \mathbf{a}$

$\overrightarrow{PG} = \mathbf{b}$

(a) Express $\overrightarrow{OM}$ in terms of $\mathbf{a}$ and $\mathbf{b}$.

$\overrightarrow{OG}$

a + b

$\overrightarrow{OM} = $ ....

(2)

(b) Express $\overrightarrow{PM}$ in terms of $\mathbf{a}$ and $\mathbf{b}$
Give your answer in its simplest form.

$\overrightarrow{PM} = $ ....

(2)

(Total for Question 1 is 4 marks)

2

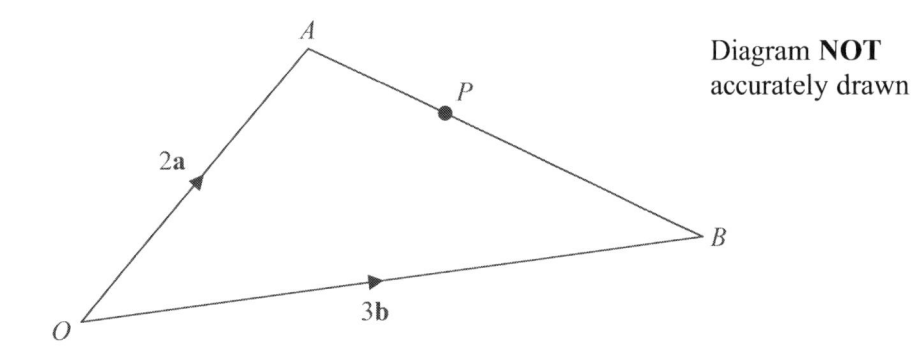

Diagram **NOT** accurately drawn

OAB is a triangle.

$\overrightarrow{OA} = 2\mathbf{a}$

$\overrightarrow{OB} = 3\mathbf{b}$

(a) Find $\overrightarrow{AB}$ in terms of **a** and **b**.

$\overrightarrow{AB} = $ −2a +3b

(1)

P is the point on *AB* such that *AP* : *PB* = 2 : 3

(b) Show that $\overrightarrow{OP}$ is parallel to the vector **a** + **b**.

(3)

(Total for Question 2 is 4 marks)

3

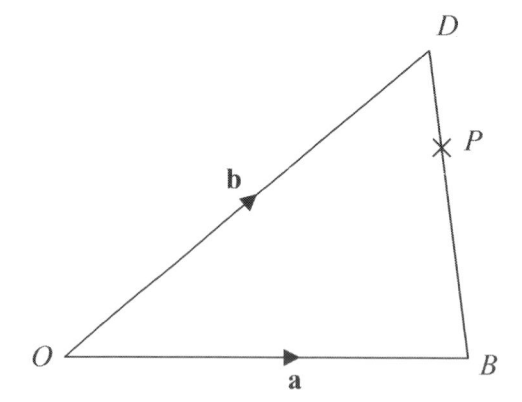

Diagram **NOT** accurately drawn

ODB is a triangle.

$\overrightarrow{OB} = \mathbf{a}$

$\overrightarrow{OD} = \mathbf{b}$

(a) Find $\overrightarrow{BD}$ in terms of **a** and **b**.

....
(1)

P is the point on DB such that $DP : PB = 1 : 3$

(b) Find $\overrightarrow{OP}$ in terms of **a** and **b**.

Give your answer in its simplest form.

....
(3)

(Total for Question 3 is 4 marks)

4

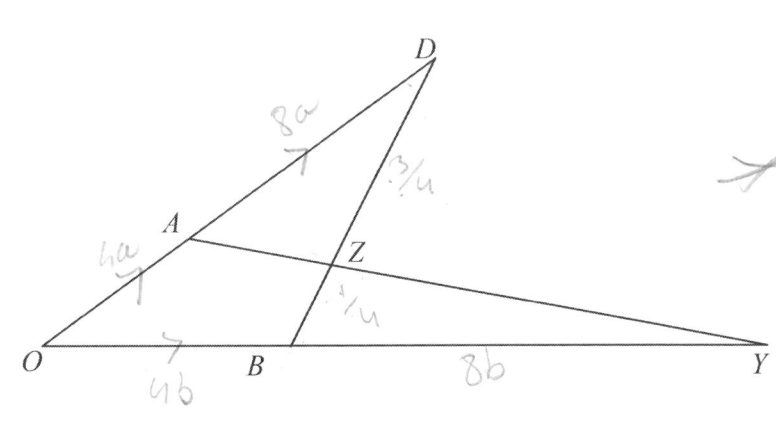

Diagram **NOT** accurately drawn

In the diagram,

$\vec{OA} = 4\mathbf{a}$ and $\vec{OB} = 4\mathbf{b}$

OAD, *OBY* and *BZD* are all straight lines

$AD = 2OA$ and $BZ:ZD = 1:3$

(a) Find, in terms of **a** and **b**, the vectors which represent

(i) $\vec{BD}$

$\vec{BD} = \vec{BO} + \vec{OD}$
$-4b + 12a$

$-4b+12a$ (2)

(ii) $\vec{AZ}$

$\vec{AZ} = \vec{AD} + \vec{DZ}$
$8a + \frac{3}{4}(-4b - 12a)$
$8a + 3b - 9a$
$-1a + 3b$

$-1a+3b$ (2)

Given that $\vec{BY} = 8\mathbf{b}$

(b) Show that *AZY* is a straight line.

$\vec{AZ} = 3b - a$

$\vec{AY} = -4a + 12b = 4(a + 3b)$

(3)

$3b - a$ same direction vector ∴ parallel, share point Z ∴ same straight line

(Total for Question 4 is 7 marks)

5

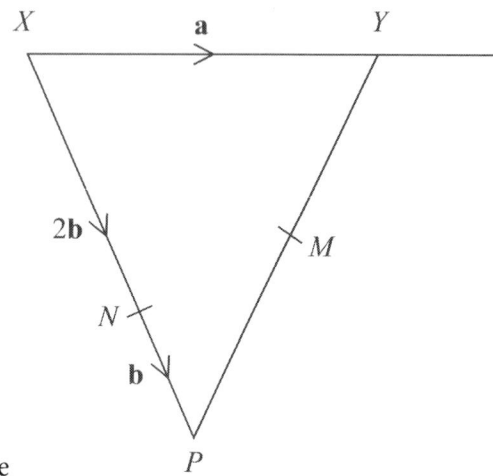

Diagram NOT accurately drawn

XYP is a triangle
N is a point on XP

$$\vec{XY} = \mathbf{a} \qquad \vec{XN} = 2\mathbf{b} \qquad \vec{NP} = \mathbf{b}$$

(a) Find the vector $\vec{PX}$, in terms of **a** and **b**.

....

(1)

Y is the midpoint of XZ
M is the midpoint of PY

(b) Show that NMZ is a straight line.

(4)

(Total for Question 5 is 5 marks)

6

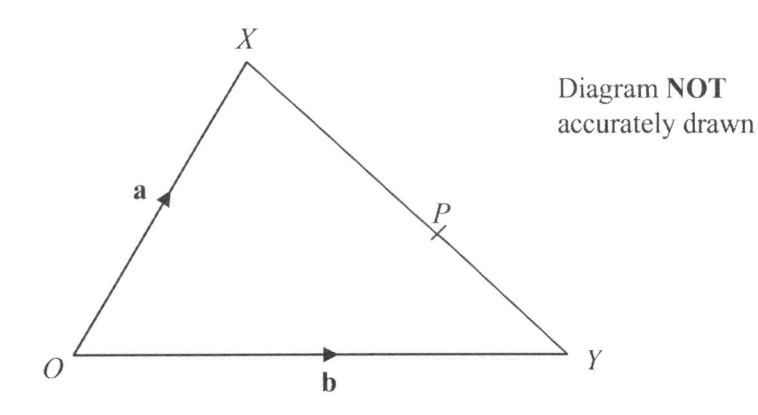

Diagram **NOT** accurately drawn

OXY is a triangle.

$\overrightarrow{OX} = \mathbf{a}$

$\overrightarrow{OY} = \mathbf{b}$

(a) Find the vector $\overrightarrow{XY}$ in terms of **a** and **b**.

$\overrightarrow{XY} = $ ....

(1)

P is the point on XY such that $XP : PY = 3 : 2$

(b) Show that $\overrightarrow{OP} = \dfrac{1}{5}(2\mathbf{a} + 3\mathbf{b})$

(3)

(Total for Question 6 is 4 marks)

7

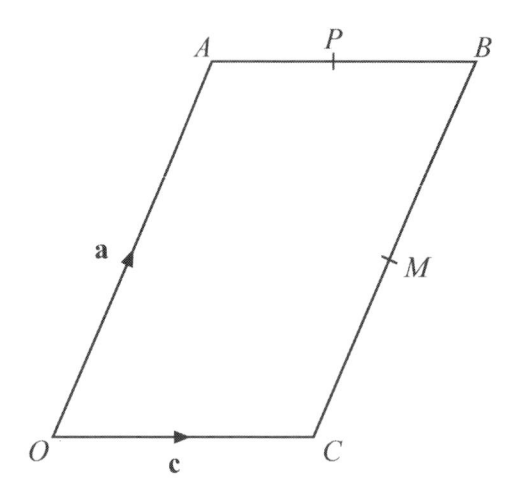

*Diagram **NOT** accurately drawn*

OABC is a parallelogram.
M is the midpoint of *CB*.
P is the midpoint of *AB*.

$\vec{OA} = \mathbf{a}$

$\vec{OC} = \mathbf{c}$

(a) Find, in terms of **a** and/or **c**, the vectors

   (i) $\vec{MB}$,

   ....

   (ii) $\vec{MP}$.

   ....

(2)

(b) Show that *CA* is parallel to *MP*.

(2)

(Total for Question 7 is 4 marks)

8

Diagram **NOT** accurately drawn

$\vec{OA} = 2\mathbf{a} + \mathbf{b}$
$\vec{OB} = 4\mathbf{a} + 3\mathbf{b}$

(a) Express the vector $\vec{AB}$ in terms of **a** and **b**

Give your answer in its simplest form.

....
(2)

Diagram NOT accurately drawn

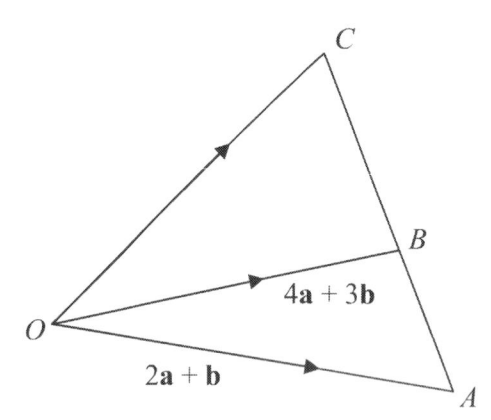

ABC is a straight line.
CB : *YZ* = 2 : 3

(b) Express the vector $\overrightarrow{OC}$ in terms of **a** and **b**
Give your answer in its simplest form.

....
(3)

(Total for Question 8 is 5 marks)

9

WXYZ is a parallelogram.
WX is parallel to *ZY*. *WZ* is parallel to *XY*.

$\overrightarrow{WX} = \mathbf{p}$

$\overrightarrow{WZ} = \mathbf{q}$

(a) Express, in terms of **p** and **q**

  (i) $\overrightarrow{WY}$

  (i) ....

  (ii) $\overrightarrow{XZ}$

  (ii) ....

(2)

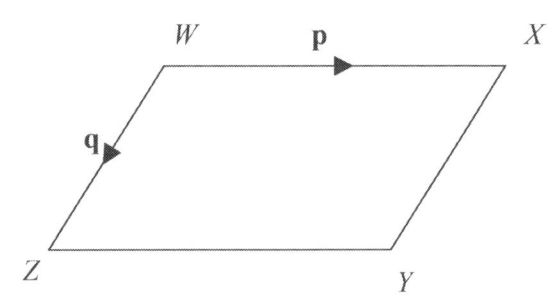

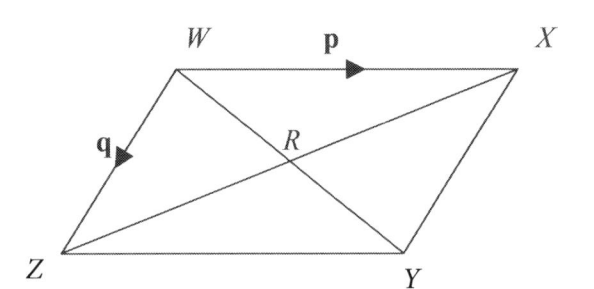

Diagram **NOT** accurately drawn

WX and *XZ* are diagonals of parallelogram *WXYZ*.
WY and *XZ* intersect at *R*

(b) Express $\overrightarrow{WR}$ in terms of **p** and **q**.

....

(1)

(Total for Question 9 is 3 marks)

10

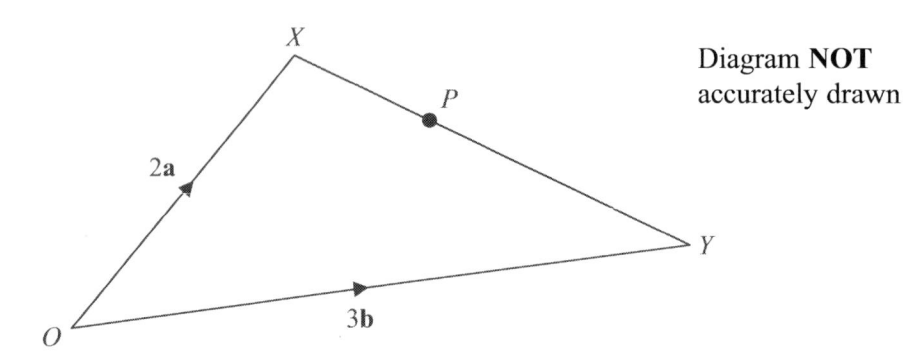

OXY is a triangle.

$\overrightarrow{OX} = 2\mathbf{a}$

$\overrightarrow{OY} = 3\mathbf{b}$

(a) Find $\overrightarrow{XY}$ in terms of **a** and **b**.

$\overrightarrow{XY} = $ ....

(1)

P is the point on *XY* such that *XP* : *PY* = 2 : 3

(b) Show that $\overrightarrow{OP}$ is parallel to the vector **a** + **b**

(3)

(Total for Question 10 is 4 marks)

11

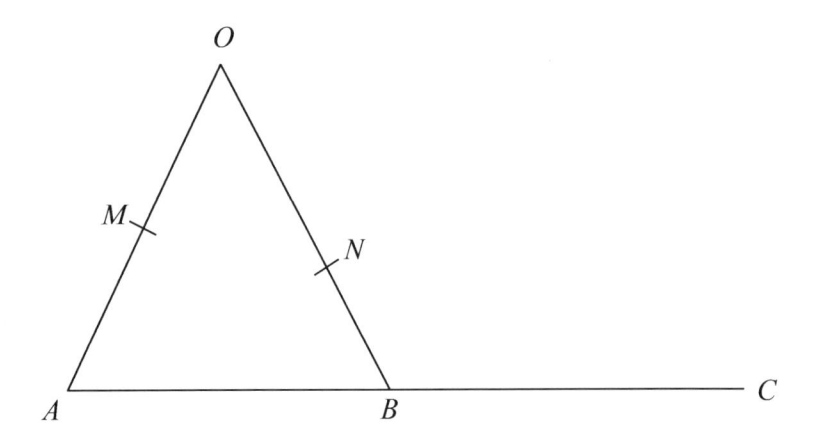

OMA, *ONB* and *ABC* are straight lines.
M is the midpoint of *OA*.
B is the midpoint of *AC*.
$\vec{OA} = 6\mathbf{a}$ $\vec{OB} = 6\mathbf{b}$ $\vec{ON} = k\mathbf{b}$ where *k* is a scalar quantity.

Given that *MNC* is a straight line, find the value of *k*.

(Total for Question 11 is 5 marks)

12 In triangle *ABC*

M is the midpoint of *AC*

N is the point on *BC* where *BN* : *NC* = 2 : 3

$\overrightarrow{AC} = 2\mathbf{a}$

$\overrightarrow{AB} = 3\mathbf{b}$

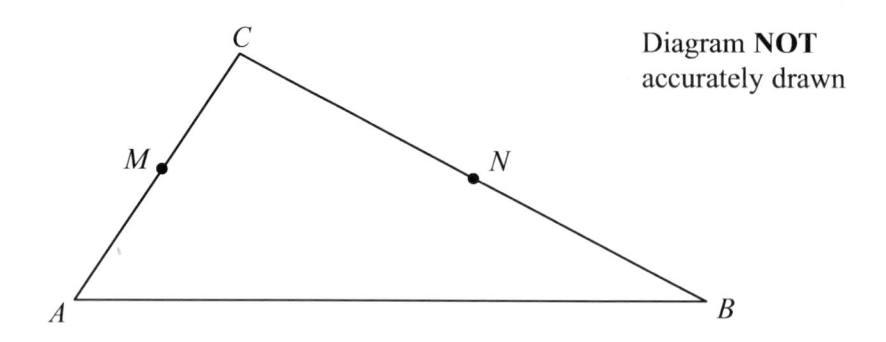

Diagram **NOT** accurately drawn

(a) Work out $\overrightarrow{MN}$ in terms of **a** and **b**.

Give your answer in its simplest form.

....

(3)

(b) Use your answer to part (a) to explain why *MN* is **not** parallel to *AB*.

(1)

(Total for Question 12 is 4 marks)